S0-AAC-474

Design Verification with *e*

Prentice Hall Modern Semiconductor Design Series

James R. Armstrong and F. Gail Gray
VHDL Design Representation and Synthesis

Jayaram Bhasker
A VHDL Primer, Third Edition

Mark D. Birnbaum
Essential Electronic Design Automation (EDA)

Eric Bogatin
Signal Integrity: Simplified

Douglas Brooks
Signal Integrity Issues and Printed Circuit Board Design

Alfred Crouch
Design-for-Test for Digital IC's and Embedded Core Systems

Daniel P. Foty
MOSFET Modeling with SPICE: Principles and Practice

Nigel Horspool and Peter Gorman
The ASIC Handbook

Howard Johnson and Martin Graham
High-Speed Digital Design: A Handbook of Black Magic

Howard Johnson and Martin Graham
High-Speed Signal Propagation: Advanced Black Magic

Farzad Nekoogar and Faranak Nekoogar
From ASICs to SOCs: A Practical Approach

Farzad Nekoogar
Timing Verification of Application-Specific Integrated Circuits (ASICs)

Samir Palnitkar
*Design Verification with **e***

Wayne Wolf
Modern VLSI Design: System-on-Chip Design, Third Edition

Kiat-Seng Yeo, Samir S. Rofail, and Wang-Ling Goh
CMOS/BiCMOS ULSI: Low Voltage, Low Power

Brian Young
Digital Signal Integrity: Modeling and Simulation with Interconnects and Packages

Design Verification with *e*

Samir Palnitkar

PH PTR

PRENTICE HALL PTR
UPPER SADDLE RIVER, NJ 07458
WWW.PHPTR.COM

Library of Congress Cataloging-in-Publication Data

Palnitkar, Samir.
    Design verification with e / Samir Palnitkar.
        p. cm.
    Includes bibliographical references and index.
    ISBN 0-13-141309-0 (alk. paper)
        1. Computer hardware description languages. 2. Integrated circuits--Verification. 3. Integrated circuits--Design & construction.
I. Title.

TK7885.7.P33 2003
621.39'2--dc22

2003058089

Editorial/Production Supervision: Wil Mara
Cover Design Director: Jerry Votta
Cover Design: Anthony Gemmellaro
Art Director: Gail Cocker-Bogusz
Manufacturing Buyer: Maura Zaldivar
Publisher: Bernard M. Goodwin
Editorial Assistant: Michelle Vincenti
Marketing Manager: Dan DePasquale

© 2004 Pearson Education, Inc.
Publishing as Prentice Hall Professional Technical Reference
Upper Saddle River, New Jersey 07458

Prentice Hall PTR offers excellent discounts on this book when ordered in quantity for bulk purchases or special sales. For more information, please contact: U.S. Corporate and Government Sales, 1-800-382-3419, corpsales@pearsontechgroup.com. For sales outside of the U.S., please contact: International Sales, 1-317-581-3793, international@pearsontechgroup.com.

Printed in the United States of America

First Printing

ISBN 0-13-141309-0

Pearson Education Ltd.
Pearson Education Australia Pty., Limited
Pearson Education Singapore, Pte. Ltd.
Pearson Education North Asia Ltd.
Pearson Education Canada, Ltd.
Pearson Educación de Mexico, S.A. de C.V.
Pearson Education—Japan
Pearson Education Malaysia, Pte. Ltd.

This book is dedicated to my family,
who provided constant support and encouragement
throughout its development.

Foreword

It is a pleasure to write a foreword to Samir's book. It tackles a really important topic: I find functional verification, in its most general sense, to be one of the most challenging and interesting engineering endeavors. It is a tough challenge, and also a critical challenge. Design bugs can kill projects, careers, and even people. In addition to being tough, verification is very creative work, and it can also be fun. I hope this book gives a glimpse of all of these aspects.

As Samir says in his preface, e is only a tool. To be successful at functional verification, you need the right tools and methodologies. You also need a lot of imagination and flexibility. Every verification project is different, because every design is different.

Much of functional verification is about knowledge representation. It is about taking the knowledge embedded in the various specifications (and in the implementation, and in people's heads), and representing it in a way that will be conducive to creating the four main components of verification: input generation, checking, coverage, and debugging.

While verification deals with highly abstract concepts, it also needs low-level operations such as Perl-style file and regular-expression handling. Thus, readers of this book may notice that e has borrowed freely from many domains to create one unified language. For instance, it contains:

- Constructs for knowledge representation (e.g., constraints and when inheritance)
- Support for Aspect Oriented Programming (AOP), for combining distinct specification aspects
- The capability to use both declarative and procedural code, including the capability to define new declarative constructs
- Low-level file and string constructs
- Constructs for describing timing, interfacing to hardware description languages (HDLs), and much more

I hope readers will find the resulting language to be coherent and natural. It has existed (in more or less its current form) for more than 10 years, and people have found more and more uses for it, from the block level to the full system level.

Lately, we at Verisity (working with many of our customers) have concentrated even more on methodology, trying to distill best-of-class methodology for reuse (culminating in the e Reuse Methodology, or eRM), and for coverage-based verification.

I would like to give special thanks to Amos Noy, Yaron Kashai, Guy Mosenson, and Ziv Binyamini, and to the many other people from Verisity who contributed to the evolution of e. Also, the language would not be what it is today without the deep involvement, helpful comments and criticism from our customers.

Finally, I'd like to thank Samir Palnitkar for writing this book. I hope he is as successful in teaching the e language via this book as he has been in teaching e in training classes.

Yoav Hollander

Founder and CTO, Verisity Design, Inc.

April 2003

Preface

During my earliest experience with *e*, I was looking for a book that could give me a "jump start" on using *e*. I wanted to learn basic digital verification paradigms and the necessary *e* constructs that would help me verify small digital designs. After I had gained some experience with building basic *e* verification environments, I wanted to learn to use *e* to verify large designs. At that time I was searching for a book that broadly discussed advanced *e*-based design verification concepts and *real* design verification methodologies. Finally, when I had gained enough experience with design verification of many multi-million gate chips using *e*, I felt the need for an *e* book that would act as a handy reference. I realized that my needs were different at different stages of my design verification maturity. A desire to fill these needs has led to the publication of this book.

Rapid changes have occurred during the past few years. High-Level Verification Languages (HVLs) such as *e* have become a necessity for verification environments. I have seen state-of-the-art verification methodologies and tools evolve to a high level of maturity. I have also applied these verification methodologies to a wide variety of multi-million gate ASICs that I have successfully completed during this period. I hope to use these experiences to make this edition a richer learning experience for the reader.

This book emphasizes breadth rather than depth. The book imparts to the reader a working knowledge of a broad variety of *e*-based topics, thus giving the reader a global understanding of *e*-based design verification. The book leaves the in-depth coverage of each topic to the reference manuals and usage guides for *e*.

This book should be classified not only as an *e* book but, more generally, as a design verification book. It important to realize that *e* is only a tool used in design verification. It is the means to an end—*the digital IC chip*. Therefore, this book stresses the practical verification perspective more than the mere language aspects of *e*. With HVL-based design verification having become a necessity, no verification engineer can afford to ignore popular HVLs such as *e*.

Currently, *Specman Elite* by Verisity Design, Inc., is the only tool that supports *e*. However, the powerful constructs that *e* provides for design verification make it an excellent HVL. Because of its popularity, it is likely that *e* will be standardized in the future and multiple vendors will create tools to support *e*. Therefore, in this book, although Specman Elite is used as a reference tool, the treatment of *e* is done in a tool-independent manner. *e* concepts introduced in this book will be generally applicable in the future regardless of the tool that is used.

Who Should Use This Book

This book is intended primarily for beginners and intermediate-level *e* users. However, for advanced *e* users, the broad coverage of topics makes it an excellent reference book to be used in conjunction with the manuals and training materials of *e*-based products.

The book presents a logical progression of *e*-based topics. It starts with the basics, such as functional verification methodologies, and *e* fundamentals, and then it gradually builds on to bigger examples and eventually reaches advanced topics, such as coverage-driven functional verification, reusable verification components, and C/C++ Interface. Thus, the book is useful to *e* users with varying levels of expertise as explained below.

Students in verification courses at universities

Parts 1, 2, and 3 of this book are ideal for a foundation semester course in *e*-based design verification. Students are exposed to functional verification methodologies and *e* basics, and, finally, they build a complete verification system with *e*.

New *e* users in the industry

Companies are rapidly moving to *e*-based verification. Parts 1, 2, and 3 of this book constitute a perfect jump start for engineers who want to orient their skills toward HVL-based verification.

Basic *e* users who need to understand advanced concepts

Part 4 of this book discusses advanced concepts such as coverage-driven functional verification, reusable verification components, and C/C++ Interface. A complete verification system example is discussed in Part 3. These topics are necessary to graduate from smaller to larger *e*-based verification environments.

e Experts

Many *e* topics are covered, from the *e* basics to advanced topics like coverage-driven functional verification, reusable verification components, and C/C++ Interface. Plenty of examples are provided. A complete verification system example is discussed in Part 3. For *e* experts, this book is a handy guide to be used along with the reference manuals.

How This Book is Organized

This book is organized into five parts.

Part 1, Introduction, presents the basic concepts of functional verification to the user. It also explains why it is important to maximize verification productivity and the methodologies used to successfully verify a digital ASIC. Finally, it discusses how an environment can be modeled using *e* for effective verification. Part 1 contains two chapters.

Part 2, *e* Basics, discusses the *e* syntax necessary to build a complete verification system. Topics covered are basics such as struct/units, generation, procedural flow control, time consuming methods, temporal expressions, checking, and coverage. This section ends with a chapter that puts together all the basic concepts and explains how to run a complete simulation with an *e*-based environment. Part 2 contains nine chapters.

Part 3, Creating a Complete Verification System with *e*, takes the reader through the complete verification process of a simple router design. Topics discussed are design specification, verification components, verification plan and test plan. The section ends with an explanation of the actual *e* code for each component required for the verification of the router design. Part 3 contains two chapters.

Part 4, Advanced Verification Techniques with *e*, discusses important advanced concepts such as coverage driven functional verification, reusable verification components (*e*VCs) and integration with C/C++. Part 4 contains three chapters.

Part 5, Appendices, discusses important additional topics such as the *e* Quick Reference Guide and interesting *e* Tidbits. Part 5 contains two appendices.

Conventions Used in This Book

The table below describes the type changes and symbols used in this book.

Visual Cue	Represents
courier	The Courier font indicates *e* or HDL code. For example, the following line indicates *e* code: `keep opcode in [ADD, ADDI];`
courier bold	In examples that show commands and their results, Courier bold indicates the commands. For example, the following line shows the usage of the Specman Elite command, **load**: `Specman>` **`load test1`**
bold	The bold font indicates Specman Elite keywords in descriptive text. For example, the following sentence contains two keywords: Use the **verilog trace** statement to identify Specman Elite events you want to view in a waveform viewer.
italic	The italic font represents user-defined variables that you must provide. For example, the following line instructs you to type the "write cover" as it appears, and then the actual name of a file: **write cover** *filename*
[ ] square brackets	Square brackets indicate optional parameters. For example, in the following construct the keywords "list of" are optional: **var** *name*: [**list of**] *type*
[ ] bold brackets	Bold square brackets are required. For example, in the following construct you must type the bold square brackets as they appear: **extend** *enum-type-name*: [*name*,...]

Visual Cue	Represents
construct, ...	An item followed by a separator (usually a comma or a semicolon) and an ellipsis is an abbreviation for a list of elements of the specified type. For example, the following line means you can type a list of zero or more names separated by commas. **extend** *enum-type-name*: [*name*,...]
|	The | character indicates alternative syntax or parameters. For example, the following line indicates that either the **bits** or **bytes** keyword should be used: **type** *scalar-type* (**bits** | **bytes**: *num*)
%	Denotes the UNIX prompt.
C1>, C2>, ...	Denotes the Verilog simulator prompt.
>	Denotes the VHDL simulator prompt.
Specman>	Denotes the Specman Elite prompt.

A few other conventions need to be clarified.

1. The words *verification engineer* and *design verification engineer* are used interchangeably in the book. They refer to the person performing the design verification tasks.

2. The words *design engineer* and *designer* are used interchangeably in the book. They refer to the person performing the logic design task. Often a logic designer also performs the design verification task.

Acknowledgements

This book was written with the help of a great many people who contributed their energies to the project.

I would like to start with a special acknowledgement for the encouragement and support given by my wife Anuradha and my sons Aditya and Sahil, who patiently put up with my late-night and weekend writing activities, and our family members, who kept me excited about the project.

This book would not have been possible without encouragement from Michael McNamara, Bob Widman, and Moshe Gavrielov of Verisity Design. I am also very honored to have Yoav Hollander, the inventor of the e language, write the foreword for my book.

Special thanks go to the following people, who put an intense effort into reviewing large parts of the manuscript.

Sean Smith	Cisco Systems
Andrew Piziali	Verisity Design
Silas McDermott	Verisity Design
Al Scalise	Silicon Image
Bill Schubert	ST Microelectronics

Joe Bauer	Vitesse Semiconductor
Mike Kavcak	SMSC
Josh Shaw	Vitesse Semiconductor
Cliff Cummings	Sunburst Design
Ray Borms	LSI Logic
Alain Pirson	Silicon Access
Hari Kotcherlakota	Silicon Image
Karun Menon	Broadcom
Craig Domeny	Integrated Design and Verification
Mike Stellfox	Verisity Design
Avinash Agrawal	SiPackets
Sun Kumbakonam	Broadcom
Shanmuga Sundaram	Intel
Rakesh Dodeja	Intel
Will Mitchell	Correct Designs
Kevin Schott	Correct Designs
Rajlakshmi Hariharan	Neomagic
Minjae Lee	Silicon Image
Rajendra Prasad	Infineon

Jamil Mazzawi	Verisity Design
Mark Strickland	Verisity Design
Dr. Paul Jackson	Edinburgh University
Dr. Steven Levitan	Pittsburgh University

I would like to extend a very special thank you to Andrew Piziali, who reviewed the book in great detail, and to Silas McDermott, who reviewed and fine-tuned the chapter exercises.

Thanks are also due to Karen Ochoa of Verisity Design, and Steve Homer of Homer Technical Publications, for patiently withstanding my deluge of FrameMaker questions and helping me produce the book. A special thanks to Linda Stinchfield for creating most of the diagrams in the book and also for helping me with FrameMaker formatting. Ron Bow copyedited and cleaned up the book in fine detail. Many thanks to him for his efforts.

Thanks to Jennifer Bilsey and Ric Chope for finalizing the cover design and for developing the marketing strategy for the book, and to Steve Brown for helping me with the history of e.

Thanks to Chen-Ben Asher and Yaron Kashai for helping me to get reviews from university faculty.

I appreciate the knowledgeable help provided by the Prentice Hall staff, including Bernard Goodwin, Wil Mara, and Nicholas Radhuber.

Some of the material in this book was inspired by conversations, email, and suggestions from colleagues in the industry. I have credited these sources where known, but if I have overlooked anyone, please accept my apologies.

Samir Palnitkar

Silicon Valley, California

Contents

List of Figures

List of Tables

List of Examples

Introduction

1 **Overview of Functional Verification**
Evolution of Verification, Verification Automation System
with e, Benefits of e

2 **Modeling a Verification Environment in e**
Interaction between Specman Elite and the Simulator,
Structs and Instances, Components of a Verification
Environment, Verification Example

CHAPTER 1

Overview of Functional Verification

As the average gate count for designs now approaches or exceeds one million, functional verification has become the main bottleneck in the design process. Design teams spend 50% to 70% of their time verifying designs rather than creating new ones. As designs grow more complex, the verification problems increase exponentially — when a design doubles in size, the verification effort can easily quadruple.

Unlike some design tasks, which have been automated with sophisticated logic synthesis or place-and-route tools, functional verification has remained a largely manual process. To eliminate the verification bottleneck, verification engineers have tried incorporating new methodologies and technologies. While various methodologies have evolved, including formal methods, simulation is still the preferred method for verification.

High-level Verification Languages (HVLs) have emerged to solve the functional verification bottleneck. This chapter describes how e, an HVL, can solve the verification bottleneck.

Chapter Objectives

- Explain the evolution of verification.

- Describe the components of a verification automation system built with e.

- Understand some advantages of e.

1.1 The Evolution of Verification

This section presents an evolution from a pure HDL-based verification methodology to a testbench automation system. Figure 1-1 represents the order of evolution in verification environments.

Figure 1-1 Evolution of Verification

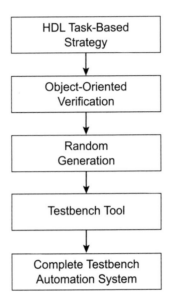

The following subsections explain these phases of evolution in greater detail. Each subsection also describes the challenges faced at each stage of evolution.

1.1.1 HDL-Based Verification

With the introduction of hardware description languages (HDLs), it became common to describe both the Device Under Test (DUT) and the test environment in VHDL or Verilog. In a typical HDL test environment:

- The testbench consisted of HDL procedures that wrote data to the DUT or read data from it.
- The tests, which called the testbench procedures in sequence to apply manually selected input stimuli to the DUT and check the results, were directed towards specific features of the design.

Figure 1-2 shows the HDL task-based strategy.

Figure 1-2 HDL Task-Based Strategy

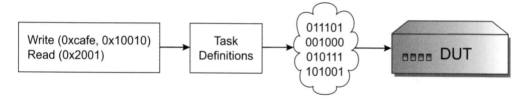

This approach broke down for large, complex designs because:

- The tests were tedious and time consuming to create.
- The tests were difficult to read and maintain.
- There were too many corner cases for the available labor.
- The environment became difficult to create and maintain because it used little shared code. HDLs did not provide robust features necessary to model complex environments.

1.1.2 Object-Oriented Verification

To make the test environment more reusable and readable, some verification engineers began to write the tests and the test environment code in an object-oriented programming language like C++.

The test writing effort was reduced, because object-oriented programming facilitated modeling the data input and output of the DUT at a high level of abstraction. The engineer created the abstract data models and let the environment create the bit-level representations of these abstract data models.

Although the amount of time spent creating individual tests was reduced, the time spent on other verification tasks increased. The test environment became more complex because new utilities, such as a simulator interface, were required. The time required to build the test environment was substantial, often overwhelming the time saved in test creation, so the overall productivity gain was not sufficient to handle the increasing design complexity.

1.1.3 Random Generation

As the efforts for object-oriented verification took root, verification engineers also realized the need to reduce the effort required to create directed tests, which were tedious, time-consuming and difficult to maintain. Therefore, verification engineers began to use random generators to automatically select input stimuli. By writing a single test and running it multiple times with different seeds, an engineer could, in effect, use the environment to create multiple tests.

However, fully random generation created a lot of illegal stimuli. In order to avoid many uninteresting or redundant tests, it was necessary to build a custom generator. Creating and maintaining a custom generator proved to be a difficult challenge.

In addition, random generation introduced new requirements:

- Checking the test results became more difficult — because the input stimuli could be different each time the test was run, explicit expected results were impossible to define before the run.

- Functional test coverage became a requirement — the engineer could not tell which tests verified which design requirements without analyzing the tests to see which input stimuli were applied.

Thus, although the test writing effort in this environment was greatly reduced, the additional work to maintain the generator meant that the overall productivity gain was not sufficient to handle the increasing design complexity.

1.1.4 Testbench Tool

At this point, there was a strong motivation to reduce the amount of effort spent in creating the complex utilities in the test environment, such as the simulator interface and the custom generator. These utilities were typically difficult to maintain when design specifications changed during the verification process. Moving to a different simulator or a different version of the design also required significant rework in the verification environment.

Typically, a testbench tool would reduce the effort required to build the test environment. However, often testbench tools did not have sophisticated constraint resolution and language-layering capability that allowed the test writer to specify the test at an abstract level, without detailed knowledge of the structure of the test environment. Therefore, the test writing effort (the most frequent activity during simulation) was still not as efficient as it could be.

In addition, testbench tools did not have a way to specify all kinds of temporal sequences and relationships. Therefore, the description of the checking requirements and complex coverage scenarios was very difficult to capture in the test environment.

1.1.5 Complete Verification Automation System

Since isolated testbench tools had certain limitations, a complete *verification automation* system that has various enabling technologies was needed to produce a significant boost in productivity. The verification automation environment needed to have the following characteristics:

- A language that allows objects in the verification environment to be extended for a particular test rather than be derived by inheritance enables a 90% reduction in test writing labor.

- A language to express constraints in a verification environment because a constraint-based approach is more powerful for testbench description. A sophisticated constraint solver and generator are needed to solve constraints between items in different objects.

- A coverage engine that allows goals to be defined for complex test scenarios.

- A temporal engine that lets the engineer capture protocol rules in a concise, declarative syntax.

A complete verification automation system increased the overall productivity of a verification environment by helping the engineer efficiently perform the following tasks:

- Defining a test plan
- Writing and maintaining the testbench environment
- Selecting test vectors (input stimuli)
- Checking results
- Measuring progress against the test plan (coverage)

The following section describes how *e* can be used to build a complete verification automation system.

1.2 Verification Automation System with *e*

The *e* language contains all constructs necessary to support the activities of a complete verification automation system. Currently, the *Specman Elite* tool from Verisity Design[1] supports the *e* language. Other tools may support *e* in the future. Figure 1-3 shows the activities in a typical verification environment.

Figure 1-3 Activities in a Verification Environment

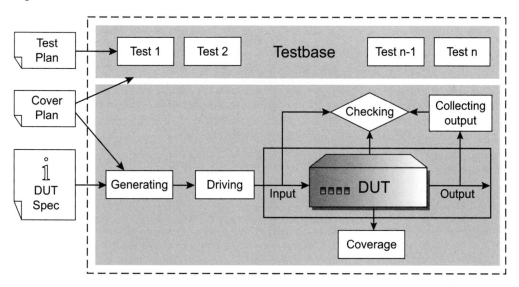

The following subsections examine how *e* assists in each verification activity.

---

1.   Specman Elite is a tool that simulates *e* code standalone or in tandem with an HDL simulator. Please see http://www.verisity.com for details on Specman Elite.

1.2.1 Generation

e automates the generation of stimuli. Test vectors are generated based on the constraints provided by the verification engineer. These constraints are typically derived from the functional and design specifications of the device and help define legal values for stimulus. Further constraints are derived from the test plan and help define specific tests. Figure 1-4 shows how tests are generated based on constraints. When a DUT is presented with an illegally constructed data stimulus, it will typically discard it without involving most of the logic in the design. Our verification goal is to exercise the entire design. Therefore, we need to make sure that most data stimulus items meet the requirements to "cross the wall" to the majority of the logic. These requirements are expressed as necessary relationships between object field values, or in other words as *constraints*.

Figure 1-4 Constraint-Based Generation

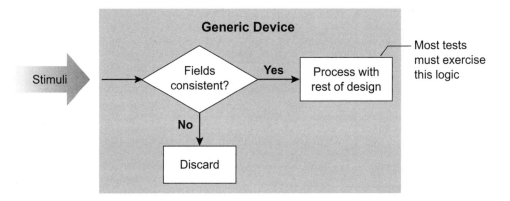

1.2.2 Driving Stimulus

After the test vectors are automatically created in the generation phase described in the previous section, these test vectors must be driven on to the DUT. e allows the drivers in a verification environment to be created in an object oriented manner. This allows maximum reusability. e syntax contains constructs to perform the following functions:

- Interface with the simulator (Verilog, VHDL, SystemC, proprietary HDL simulator, hardware accelerator, or in-circuit emulator[2])
- Coordinate all procedural code that drives the DUT

---

2.   Though e is capable of interfacing with any software or hardware simulator, Verilog and VHDL simulators are covered in this book.

- Conform to input protocol for DUT
- Convert abstract data structures into a list of bits
- Drive the bits onto the DUT

Figure 1-5 shows the components provided by *e* for driving the DUT.

Figure 1-5 Driving the Stimulus on to the DUT

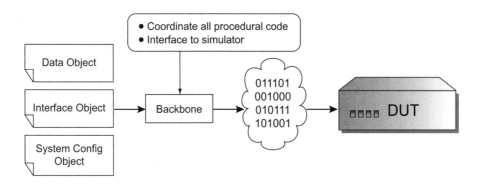

1.2.3 Collecting Output

After the stimulus is injected into the DUT, output is produced from the DUT. This output must be collected and checked. *e* allows the receivers in the verification environment to be created in an object oriented manner. This allows maximum reusability. *e* syntax contains constructs to perform the following functions:

- Interface with the simulator (Verilog or VHDL or SystemC or proprietary HDL simulator or hardware accelerator or in-circuit emulator)
- Recognize the output protocol for DUT
- Receive the bits from the DUT
- Coordinate all procedural code that receives data from the DUT
- Convert a list of bits into an abstract data structure for checking

1.2.4 Data Checking

After the output data are received from the DUT, the data must be checked. There are two types of checks that need to be performed.

Data Value Checking   The list of bits received from the DUT must be compared against the expected data. To simplify the comparison, *e* has constructs that convert a list of bits to an abstract data structure. *e* also has constructs to aid in the comparison of abstract data structures.

The fields of the expected data structure and the output data structure are compared to determine success or failure.

Temporal Checking Temporal checks simply mean "timing" checks. These checks ensure that protocol is being followed at each interface of the DUT. For example, if the protocol requires that a grant should follow a request within five clock cycles, this requirement is handled by temporal assertions. Temporal checking constructs are used to build protocol monitors.

Do not confuse temporal checks with the timing analysis of logic circuits, setup and hold delays, etc. Temporal checks are functional/protocol checks, whereas timing analysis of logic circuits contains checks in terms of setup and hold times.

Figure 1-6 shows two types of checks that need to be performed on the data.

Figure 1-6 Data Checking

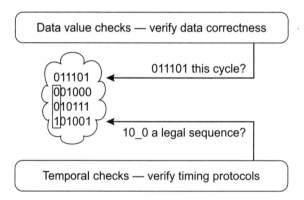

The data checking method described in this section is just one example of a verification automation system. Another method that is very prevalent in central processing unit (CPU) verification is to use a reference model to check the output of the DUT. Both the CPU and the reference model are simulated in parallel and their outputs and register states are compared for equality.

1.2.5 Coverage Measurement

Functional coverage results tell the verification engineer if the test plan goals have been met. There are three types of functional coverage:

Basic Item Coverage This coverage tells the engineer if all legal values of an interesting variable have been covered.

Transition Item Coverage This coverage is used for state machines which form the control logic for any design. This coverage tells the engineer if all legal transitions of a state machine have been covered.

Cross Coverage This coverage allows the engineer to examine the cross product of two or more basic or transition items to check if all interesting combinations of basic and transition items have been covered.

e contains constructs to specify all the above coverage items.

Code coverage is usually performed in addition to functional coverage. *e* provides a way of connecting to code coverage tools.

1.3 Benefits of *e*

The primary objective of any verification automation system is to increase productivity and DUT quality. Some benefits of an HVL such as *e* are listed below:

- Environments can be built with declarative code (instead of procedural code that executes sequentially). This makes the environment easy to extend and modify. This also minimizes the amount of procedural code that needs to be written.
- *e* provides dynamic inheritance that allows the environment to be built in an object oriented manner.
- *e* has constructs for constraint-driven generation, temporal language, and functional coverage. These constructs help the engineer quickly build various components in the environment.

There are many other benefits of *e*. These benefits will become obvious in later chapters as we discuss *e* in detail.

1.4 Summary

Verification methodologies have evolved over the years from a simple HDL task-based methodology to a complete verification automation system. With designs today reaching a few million gates, verification automation systems are now required to maximize productivity of the verification process.

e provides all constructs needed to build an integrated verification automation system in an object oriented manner and an aspect oriented manner. Currently, the *Specman Elite* tool from Verisity Design supports the *e* language. Other tools may support *e* in the future.

e can be used to construct components to do the following functions in a verification environment:

Generation *e* automates the generation of stimuli. Input stimuli are generated based on the constraints provided by the verification engineer.

Driving Stimulus After the test vectors are generated, they must be driven on to the DUT. *e* provides a simulator interface and the necessary mechanism to drive the DUT.

Collecting Output   After the stimulus is applied to the DUT, output is produced from the DUT. This output must be collected and checked. *e* provides a simulator interface and the necessary mechanism to receive data from the DUT.

Data Checking   After the output data are received from the DUT, the data must be checked. Data value checks compare the output data values against the expected data. Temporal assertions monitor the functional protocol at important interfaces.

Coverage   Functional coverage results tell the verification engineer if the test plan goals have been met. There are three types of coverage: basic item coverage, transition item coverage, and cross coverage.

Modeling a Verification Environment in *e*

Before we discuss the details of the *e* language, it is important to understand the various components of a verification environment built with *e*. A verification engineer must use "good" techniques to do efficient verification. In this chapter we discuss the high level components of a verification environment and the interaction between them. There are *e* code examples in this chapter to illustrate the high level components. The syntax details of these examples will be discussed in later chapters. In this chapter, readers should focus only on the high level concepts.

Currently, the *Specman Elite* tool from Verisity Design[1] supports the *e* language. We will use this tool as a reference *e* implementation.[2] We will discuss the interaction between *Specman Elite* and the *Verilog or VHDL simulator*.[3] Note that although Specman Elite is used, the concepts explained in this book will apply to any future tools that support the *e* language.

Chapter Objectives

- Understand interaction between Specman Elite and the Simulator.
- Explain the importance of structs and units.
- Describe components of a good verification environment.
- Understand a simple verification system designed with *e*.

---

1. Please see http://www.verisity.com for details on Specman Elite.

2. Other tools may support *e* in the future.

3. For conciseness, in this book, hereafter we usually will refer to the Verilog or VHDL simulator as simply *Simulator* or *HDL Simulator*.

2.1   Interaction between Specman Elite and the Simulator

Specman Elite and the Simulator are two separate processes that run concurrently and synchronize with each other during the simulation. Specman Elite and the Simulator talk to each other through an interface that includes a special file called a *stubs* file. The stubs file provides the primary interface for communication between Specman Elite and the Simulator. In addition to the stubs file, Specman Elite and the Simulator also communicate through mechanisms such as Verilog Programming Language Interface (PLI) or the VHDL Foreign Language Interface (FLI). Figure 2-1 shows this interaction.

Figure 2-1 Interaction Between Specman Elite and the Simulator

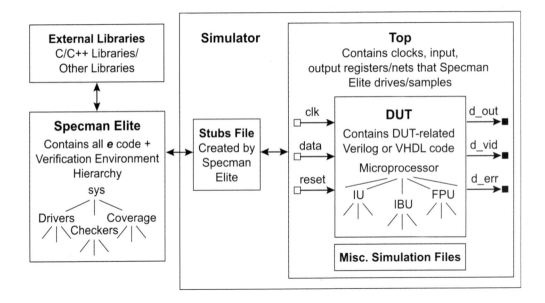

2.1.1   Components of Specman Elite and the Simulator

There are multiple components in a simulation environment containing *e*.

Specman Elite   The entire verification environment is modeled in *e*. This environment contains driver, checkers, coverage objects, data objects, etc., required for the verification of the DUT. All *e* files are compiled and simulated by Specman Elite.

External Libraries   If any legacy code or models exist in C/C++ or other languages, they can be invoked from *e* code. For example, if an Motion Picture Experts Group (MPEG) decoder algorithm is already coded in C++, it is easy for *e* code to simply call that algorithm to make comparisons against expected frames. Such external libraries may or may not be present in an *e* based simulation.

DUT The Device Under Test contains the entire hierarchy for the design. For example, if a microprocessor is coded in Verilog, it will have various modules such as Integer Unit (IU), Interface Bus Unit (IBU), Floating Point Unit (FPU), etc. The DUT represents the topmost level of the design hierarchy and is typically written entirely in Verilog or VHDL.

Miscellaneous Simulation Files These are Verilog or VHDL files that may represent legacy models, bus functional models, etc., required for verification of the DUT. For complex environments, it is recommended that all verification files be coded in *e*. In these cases, such miscellaneous Verilog or VHDL files will not be present in an *e* based simulation. However, initially, the verification engineer may choose to leave some legacy verification modules in Verilog or VHDL. Therefore, they need to be included in the simulation.

Top This module represents the top level of the simulation hierarchy. All files in this hierarchy are coded in Verilog or VHDL. The top module instantiates the DUT and other miscellaneous verification modules. The top module also contains registers that drive the inputs of the DUT and nets that receive the outputs of the DUT. In Figure 2-1, the empty squares represent registers and dark squares represent nets.

Stubs File This is a special Verilog or VHDL file that is automatically created by Specman Elite when a special command is issued by the verification engineer. This file acts as the communication channel between Specman Elite and the Simulator during simulation. The process of creating this file is very simple. All *e* files are read into Specman Elite. Then, a special Specman Elite command is issued to create this file. The Simulator is not required to be present to create this file. This stubs file is then included in the list of Verilog or VHDL files to be compiled by the Simulator during the simulation. This file needs to be created very infrequently.

Simulator This is a Verilog or a VHDL simulator that runs concurrently with Specman Elite. The Simulator compiles and simulates all Verilog or VHDL files. At the top-most level, there are two modules—the *top* module and the module inside the stubs file. The Simulator and Specman Elite communicate through the stubs file.

2.1.2 Flow of Simulation

The following steps summarize the flow of execution between Specman Elite and the Simulator. Although we have not talked about the details of the *e* language syntax or Specman Elite, these steps will help you understand the basics of the simulation execution flow.

1. Both Specman Elite and the Simulator are invoked simultaneously. Specman Elite gains control and the Simulator stops at simulation time t=0. The control is explicitly passed to the Simulator and the simulation is started.

2. The Simulator continues to run until it reaches a trigger point set by Specman Elite on the change in value of a Simulator variable. This trigger point or *callback* is set by the verification engineer inside the *e* code. At that callback, the Simulator halts further execution and transfers control to Specman Elite.

3. Specman Elite does all necessary computations. When all computations are waiting for the next trigger point, Specman Elite has nothing more to do, so it transfers control back to the Simulator. During this entire process, the Simulator is not executing further. Therefore, no simulation time has elapsed.

4. Simulator continues simulation until it reaches the next trigger point. Then it repeats steps 2 and 3.

5. Simulation continues with control passing back and forth between Specman Elite and the Simulator until a procedure inside Specman Elite requests to finish the simulation run.

6. The results may then analyzed. Based on the results a new run can be started.

The following section describes how a verification hierarchy is created using *e*.

2.2 Structs and Instances

Figure 2-1 shows that the *e* hierarchy and the design hierarchy (Verilog or VHDL) are two independent trees. The *e* hierarchy is focused on creating verification components. The Verilog or VHDL hierarchy represents the logic design components. For example, a Verilog design hierarchy consists of modules and module instances. Similarly, the *e* hierarchy consists of **struct** and **struct** instances. Figure 2-2 shows an example of *e* and Verilog hierarchies in the same simulation run.

Figure 2-2 *e* and Verilog Hierarchies

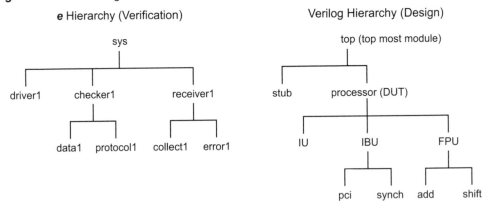

As shown in Figure 2-2, the top of the *e* verification hierarchy is a **struct** called **sys**.[4] This is an implicitly defined **struct** in Specman Elite. In Verilog, the top level module can be given any

---

4.   In reality, **sys** is a **unit**. We have yet not introduced the concepts of units, so we refer to **sys** as a struct. Moreover, there are other structs above **sys**. However, from a verification engineer's perspective, **sys** represents the tree where the verification environment is built.

name and it needs to defined explicitly. In *e*, the top level **struct** is always **sys**, and it is always defined implicitly. However, all other structs are instantiated explicitly inside **sys**. All structs such as driver1, checker1, and receiver1 that are defined in the *e* environment must be instantiated in the **sys** tree structure just as design modules such as IU, IBU, FPU need to be instantiated in the tree structure of the DUT which is located under the top level Verilog module. Example 2-1 shows how the described hierarchy is built in *e* code. A **struct** is defined for each verification component. Then the **struct** is instantiated under higher level structs. Finally, the highest level structs are instantiated under **sys**.

Example 2-1 Building Hierarchy under sys

```
<' //Indicates starting of e code
struct data { //Define the data struct. To be instantiated inside
             //checker
--
<data internals>
--
};
struct protocol {//Define the protocol struct. To be instantiated
              //inside checker
--
<protocol internals>
--
};

struct collect { //Define the collect struct. To be instantiated
              //inside receiver
--
<collect internals>
--
};

struct error {   //Define the error struct. To be instantiated
              //inside receiver
--
<error internals>
--
};

struct driver {
--
<driver internals>
--
};
```

Example 2-1 Building Hierarchy under sys (Continued)

```
struct checker {
data1: data; // Instantiate the struct data and call it data1
protocol1: protocol; // Instantiate the struct protocol and
                     // call it protocol1
--
<checker internals>
--
};
struct receiver {
collect1: collect; // Instantiate the struct collect and
                   // call it collect1
error1: error; // Instantiate the struct error and
               // call it error1
--
<receiver internals>
--
};

extend sys { //sys is implicitly defined. So extend it to add
            //new instantiations. This is the topmost struct
            //for the verification environment.
driver1: driver;
checker1: checker;
receiver1: receiver;
};
//End of e code
'>
```

As shown in Example 2-1, the **extend** keyword is used to add additional instances to **sys**. Similarly, the **extend** keyword can be used to enhance the definition of any user-defined struct. Thus *e* lends itself very well to a layered development approach. The functionality for the base structs can be extracted from the design specification. Any test-specific changes can then be added separately using the **extend** mechanism for the structs.

The following section explains the various components in a typical verification environment.

2.3 Components of a Verification Environment

A typical *e* based verification environment contains many different components. Figure 2-3 shows these components.

Figure 2-3 Components of a Verification Environment

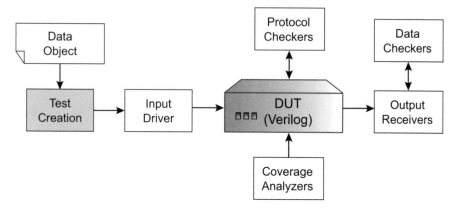

Each component shown in Figure 2-3 above is defined as a **struct** (or a **unit** as discussed in later chapters).

2.3.1 Data Object

A **struct** is defined to represent a data object. This represents one test vector or one stimulus item. The data object contains fields that define the stimulus item. Different tests have variations in values of these fields. For example, in a router, the stimulus item could be a single packet. In a video application, the stimulus item could be a single frame. In a microprocessor, the stimulus item could be a single instruction. Specman Elite contains a random stimulus item generator that can automatically generate the value of a stimulus item based on certain constraints.

2.3.2 Test Creation

Tests are simply a set of constraints placed on the generation of the fields in the data object. If there are no constraints, the fields are selected randomly. If the constraints are very restrictive, the test becomes more and more directed. In *e*, there are no long task-based procedural tests that are very difficult to maintain.

2.3.3 Input Driver

A **unit**[5] is defined to represent a driver object. The driver object contains an input procedure to take one stimulus item and apply it to the DUT. The procedure must follow the interface protocol expected by the DUT at the input ports. In addition, the driver object also has a master procedure that calls the input procedure multiple times to apply many stimulus items to the DUT. Depending on the number of input interfaces, there may be multiple instances of the input driver.

---

5.   A **unit** is very similar to a **struct** but has some additional capabilities. A **unit** is used to define a static verification object that does not move through the verification system. A struct is a dynamic object or a data item such as instruction or packet that moves throughout the system. A **unit** is defined in greater detail in later chapters.

2.3.4 Output Receiver

A **unit** is defined to represent a receiver object. The receiver object contains a procedure to collect raw output from the DUT and convert it to a data object format. The procedure must follow the interface protocol expected by the DUT at the output ports. The receiver then passes this data object to the data checker to compare it against expected data.

2.3.5 Data Checker

A **unit** is defined to represent a data checker object. The data checker object gets an output data object from the receiver and compares it against the expected data. The data checker has a procedure to generate and store the expected data object. The data checker may be instantiated in the receiver object depending on the environment. Alternately, the data checker may be a central object instantiated directly under the verification environment.

2.3.6 Protocol Checker

A **unit** may be defined to represent a protocol checker object. Protocol checkers are monitors built using temporal assertions. The protocol checker object normally contains a set of rules that monitor the protocol at all important input and output interfaces. If the protocol is violated, the protocol checker issues a warning or an error. The declarative syntax of *e* makes it really simple to write these protocol checkers. The protocol checker can be designed as a centralized object or a distributed object depending on the architecture of the verification environment.

2.3.7 Coverage Analyzer

A **unit** may be defined to represent a coverage object. A coverage object defines a set of basic, transition, and cross coverage items to monitor upon specific events in the simulation. Coverage statistics are gathered in the coverage object during simulation. After the simulation is finished, a summary of these statistics is displayed. This summary helps the engineer analyze the untested areas and write new tests to cover those areas. The coverage analyzer can be designed as a centralized object or a distributed object depending on the architecture of the verification environment.

2.3.8 Verification Environment

A **unit** is defined to represent a verification environment object. Typically, all the objects defined above can be directly instantiated in the **sys** structure. However, it is better to encapsulate the verification environment by instantiating all the above objects under a special verification environment object. Then the verification environment object is instantiated in the **sys** structure. This makes the entire environment reusable and portable.

Note that all static objects related to the environment are defined as **unit**. Variable objects such as stimulus items are defined as **struct**. The difference between structs and units will be covered in later chapters.

Figure 2-4 shows the hierarchy of a typical verification environment. There may be multiple instances of each **struct** or **unit** depending on the structure of the design. Note that this hierarchy is separate from the design hierarchy in Verilog or VHDL.

Figure 2-4 *e* Verification Environment Hierarchy

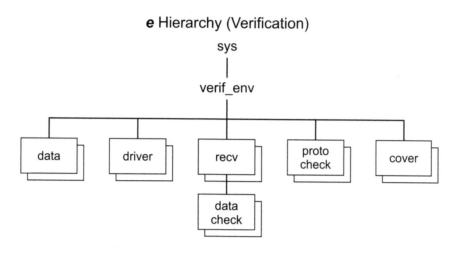

Figure 2-5 summarizes how an *e*-based environment is partitioned. Specman Elite provides the constraint solver and the backbone to simulate with the *e* language. The DUT is simulated in Verilog or VHDL. All environments contain objects such as interface objects and system objects that are static in nature, i.e., they are written once as defined by the specification. Interface objects drive the data onto the DUT interfaces. System objects contain the high-level organization of various interface objects. These objects are declared as **unit** and are written once per project. Stimulus items are dynamic data objects whose fields might be varied many times in one test. Such data objects are defined as **struct**. Multiple tests are created by constraining the data object, interface object, and system objects.

Figure 2-5 *e* Verification Environment Partitioning

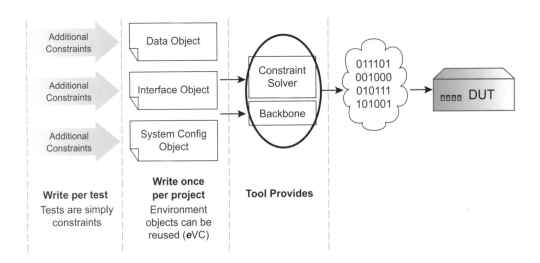

The following section discusses a small *e* example.

2.4 Verification Example

The simple XOR verification example discussed in this section does not cover all the components discussed in "Components of a Verification Environment" on page 18. However, it does give the reader an idea of building a verification environment in *e*.

The purpose of this example is to introduce the reader to *e*. Syntax details are not explained in this section but will be covered in later chapters. It is important to pay attention to only the high-level verification structure at this point. We will learn *e* syntax in great detail in later chapters. There are plenty of comments throughout the examples in this section. However, the actual *e* code is very simple.

A Verilog example is described in this section. However, *e* code will work with either a Verilog or VHDL design.

2.4.1 XOR Device Under Test (DUT) Description

The design to be tested is a simple XOR flipflop. Figure 2-6 shows the circuit diagram for this flipflop.

Figure 2-6 XOR DUT Description

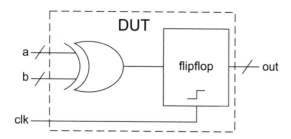

The Verilog code for the XOR DUT is shown in Example 2-2 below.

Example 2-2 Verilog Code for XOR DUT

```
//Definition of XOR Module
'define width 2
module xor_mod (out, a, b, clk);
parameter width = 'width;
input clk; // 1 bit input
input   [width-1:0] a, b; // 2 bit inputs

output [width-1:0] out; // 2 bit output
reg    [width-1:0] out;
always @(posedge clk) out = #1 (a ^ b);
endmodule
// xor module
```

2.4.2 Simulator Hierarchy

The XOR DUT is instantiated inside the module *xor_top*. Module *xor_top* instantiates the DUT, creates a clock whereby *e* code synchronizes with DUT and declares variables a, b, and *out*. The Verilog code for the *xor_top* module is shown in Example 2-3.

Example 2-3 Top Level Verilog Module

```verilog
// Top level module to test an n-bit XOR Verilog design
// that is to be tested using e
'define width 2 //Define a 2 bit width

//Define the top level module
module xor_top;
  //Define the parameters
  parameter width = 'width;
  parameter clock_period = 100;

  //Define variables for a, b and out
  reg [width-1:0] a, b;
  wire [width-1:0] out;

  // Setup clock. This clock will be used by
  // e code to synchronize with DUT.
  reg clk;

 initial clk =0;
  always #(clock_period/2) clk = ~clk;

  //Verilog monitor task
  task mon ;
    begin
      $monitor($time,"clk=%b a=%b b=%bout=%b",clk,a,b,out);
    end
  endtask

  // Instantiate the n-bit XOR DUT
  xor_mod x1(out,a,b,clk);

endmodule
```

Thus, in the simulator, the hierarchy of the Verilog design is as shown in Figure 2-7 below.

Figure 2-7 Verilog Hierarchy for XOR Example

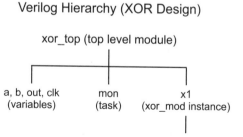

Verilog Hierarchy (XOR Design)

xor_top (top level module)

a, b, out, clk   mon   x1
(variables)   (task)   (xor_mod instance)

a, b, out
(variables)

2.4.3 Test Plan

The test plan for the above DUT is very simple.

1. Test all legal combinations of *a* and *b* (any 0,1 combination is legal).

2. Check values of *out* against expected values.

The *clk* variable will be used by *e* code to synchronize with the DUT. The test plan will be executed from *e* code, i.e., the values of inputs *a* and *b* will be chosen and toggled from within *e* code. Figure 2-8 shows the interaction between Specman Elite and the Simulator.

Figure 2-8 Interface between Specman Elite and the Simulator for XOR DUT Verification

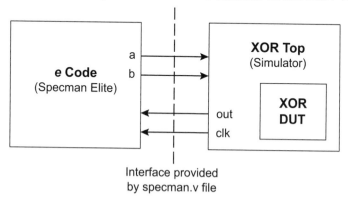

2.4.4 *e* Verification Hierarchy

Various verification components will be needed to verify the XOR DUT. The test hierarchy to verify the XOR DUT will be represented in *e*. Figure 2-9 shows a possible hierarchy for the verification environment. We have omitted the protocol checker from this example for the sake of brevity.

Figure 2-9 Verification Environment Hierarchy

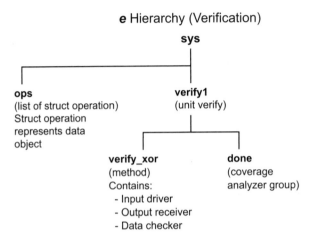

2.4.5   *e* Code for the Verification Environment

The *e* code for the verification environment is shown in Example 2-4 below. Each component in the verification is highlighted with comments. There is no need to understand the syntax for the *e* code shown in this example. The purpose of displaying the *e* code is to give the reader a feel for the syntax. Details of *e* syntax will be described in later chapters.

Example 2-4 *e* Code for Verification Environment

```
This is a test environment for the XOR DUT written in e
      out[1:0] == (a[1:0] ^ b[1:0])
Beginning of e code to verify XOR DUT
<'
//----------------------------------------------------
// Data object. This is the basic stimulus item that will
// be applied to the DUT.
struct operation {
  a: int (bits: 2 ); //input to DUT
  b: int (bits: 2 ); //input to DUT
  !result_from_dut: int (bits:2); //output from DUT
};
```

Example 2-4 *e* Code for Verification Environment (Continued)

```
//----------------------------------------------------
// sys instantiates a list of basic stimulus items. These
// form the test vectors to apply to DUT. The number
// of test vectors is constrained to be less than 20.
extend sys {
  ops: list of operation; // List of stimulus items
  keep ops.size() < 20; //Keep number of stimulus items < 20
};

//----------------------------------------------------
// Can call a Verilog task directly from within e code
verilog task 'xor_top.mon'(); //Need full path name

//----------------------------------------------------
// struct verify. Input Driver, Output Receiver and
// Data Checker functions are all combined into this
// struct. Typically, these functions would be in separate
// structs or units.
struct verify {
//Create an event fall_clk that will be triggered
//at each falling edge of clk. Specman Elite will be
//invoked by the Verilog simulator at each occurrence.
event fall_clk is fall('~/xor_top/clk')@sim;

//Define a time consuming method to run input driver,
//output receiver and data checker functions
//This procedure is synchronized to falling edge of clk.
verify_xor() @fall_clk is {
'xor_top.mon'(); //Call the Verilog task
for each operation (op) in sys.ops { //For each data item
                                     //in list execute block

//Write values to Verilog variable
//This part does task of Input Driver
'~/xor_top/a' = op.a; //Apply value of a to Verilog variable
'~/xor_top/b' = op.b; //Apply value of b to Verilog variable

wait [1]; // wait one cycle, i.e till next falling edge

//Read values from Verilog variable
//This part does task of Output Receiver
op.result_from_dut = '~/xor_top/out1';
```

Example 2-4 *e* Code for Verification Environment (Continued)

```
print op; // Print the current data object

//Check output against expected values
//This part does task of Data Checker
check that op.result_from_dut == (op.a ^ op.b);

//Emit an event for the coverage group to sample data
emit op.done;
}; //End of for each loop

//Stop the run, finish simulation
stop_run();
}; //End of method verify_xor


//Invoke the verify_xor method at simulation time 0
run() is also {
    start verify_xor();
};

}; //End of struct verify

//----------------------------------------------------
//Extend sys to instantiate the verify struct defined above
extend sys {
verify1: verify; //Instance named verify1
};

//----------------------------------------------------
//Define coverage analyzer group
//Coverage group is defined inside the struct operation
//instead of defining it as a separate struct or unit
extend operation {
    event done; //Whenever this event is emitted, take
                //a snapshot of the items below
    cover done is {
        item a; //Snapshot of value of a
        item b; //Snapshot of value of b
        cross a, b; //Cross coverage of a and b values
    };
};
```

Example 2-4 *e* Code for Verification Environment (Continued)

```
//Setup needed to enable coverage
extend sys { //Extend sys
                //to set up tool configuration.
  setup_test() is also {
    set_config(cover, mode, count_only); //Enable count_only coverage
  };
//---------------------------------------------------------


'>
End of e code to verify XOR DUT
```

2.4.6 Specman.v File

The *specman.v* file contains the interface between the Simulator and Specman Elite[6]. This file is compiled into the Verilog simulator and run with the other DUT files. This file is created *automatically* by Specman Elite. All *e* code is loaded into Specman Elite and then a command is invoked to create this file. Figure 2-10 shows the interaction between Specman Elite and the HDL Simulator.

Figure 2-10 Interaction between Specman Elite and the Simulator for XOR DUT Verification

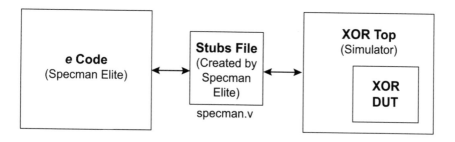

The *specman.v* file should never be modified manually. It should always be created with a Specman Elite command. Specman.v file contains a Verilog module called *specman*. Example 2-5

6.   The file *specman.vhd* is created in a VHDL environment.

shows the code generated for the *specman.v* file generated for the XOR DUT verification example. There is no need to understand the code inside the *specman.v* file.

Example 2-5 Specman.v File Code for XOR DUT Verification

```
/* specman.v - A Specman Elite stubs file (top module is xor) */
/* Generated automatically on Thu Oct 31 12:15:25 2002*/

module specman;

    parameter sn_version_id = 944493104;  /* Version */
    parameter sn_version_date = 100998;   /* Version date*/

 event sn_never_happens;
    always @sn_never_happens begin
        $sn();
    end

    reg sn_register_memory;
    event sn_register_memory_event;
    initial begin
        sn_register_memory=0;
    end
    always @(sn_register_memory_event) begin
        sn_register_memory=1;
    end
 initial begin
        ->sn_register_memory_event;
    end

    event require_specman_tick;
    event emit_specman_tick;

    always @(require_specman_tick) begin
        #0 ->emit_specman_tick ;
    end


 /* Verilog task xor_top.mon */
reg mon_0_trig;

always @(posedge mon_0_trig) begin
xor_top.mon;
mon_0_trig = 0;
->require_specman_tick;
end
```

Example 2-5 Specman.v File Code for XOR DUT Verification (Continued)

```
reg finish_test;
always @(posedge finish_test) begin
$finish;
end

endmodule /* specman */

module specman_trace;
endmodule /* specman_trace */
```

2.4.7 Running the Simulation

The following steps summarize the flow of running the simulation to verify the XOR DUT example.

1. To run the simulation, both Specman Elite and the Simulator processes are invoked simultaneously. Specman Elite gains control and the Simulator stops at simulation time t=0. The control is then explicitly passed to the Simulator and the simulation is started.

2. The Simulator continues to run until it reaches a falling edge of the clock (set by event *fall_clk* which has its sampling event **@sim**). At this point, the Simulator halts further execution and transfers control to Specman Elite.

3. Specman Elite executes the first action in the *verif_xor()* method at the first falling edge of the clock. The *verif_xor()* method applies the *a* and *b* inputs to the DUT. Then, it encounters a **wait** keyword, so it transfers control back to Simulator.

4. The Simulator continues simulation until it reaches the next falling edge of the clock. Control is passed back to Specman Elite. The *verif_xor()* method gathers the output and checks the output against the expected value. An event *done* is emitted for the coverage analyzer to take a sample. Then it goes to the top of the **for each** loop and applies the *a* and *b* inputs of the next stimulus item to the DUT.

5. Simulation continues with control passing back and forth between Specman Elite and the Simulator until *a* and *b* inputs for all stimulus items in the *ops* list are applied to the DUT and the outputs are checked. When all stimulus items are applied, Specman Elite quits the **for each** loop and encounters a **stop_run()** invocation. This indicates completion of simulation.

6. The coverage results can be displayed. The coverage report tells which values of *a* and *b* inputs have not been covered. The next test can cover these values until all possible values of *a* and *b* inputs are covered.

2.5 Summary

- Specman Elite and the Simulator are invoked concurrently and synchronize with each other during the simulation run. Specman Elite and the Simulator talk to each other through an interface that is included in a special file called a *stubs* file. The stubs file provides the channel for communication between Specman Elite and the Simulator.

- An *e* environment is made up of objects declared using **struct** and **unit** keywords. These objects form the verification hierarchy. Static objects related to the environment are defined as **unit**. Variable objects such as stimulus items are defined as **struct**.

- In *e*, the top level **struct** is always **sys**, and it is always defined implicitly. However, no structs are instantiated implicitly inside **sys**. All structs in the *e* environment must be explicitly instantiated in the **sys** tree structure. In Verilog, modules are instantiated to create hierarchy. In VHDL, entities are instantiated to create hierarchy.

- A typical *e*-based verification environment may contain objects to represent data object (stimulus item), input driver, output receiver, data checker, protocol checker, and coverage analyzer. Some objects may not be present depending on the complexity of the verification process. These objects may be defined independently or as a part of another object.

2.6 Exercises

1. Define the role of a stubs file in the interaction between Specman Elite and the Simulator. Review whether it is necessary to edit the stubs file manually?

2. Describe the flow of execution between Specman Elite and the Simulator.

3. Write the *e* code to create the following verification hierarchy. Just create all objects using the **struct** keyword. You do not need to fill in internal details of the objects. The figure shows names of instances. The names of structs are in parentheses.

e Hierarchy (Verification)

4. Describe the components of a typical verification environment.

e Basics

3 **Basic *e* Concepts**
Conventions, Data Types, Simulator Variables, Syntax Hierarchy

4 **Creating Hierarchy with Structs and Units**
Defining Structs, Extending Structs, Defining Fields, Defining List Fields, Creating Struct Subtypes with when, Units

5 **Constraining Generation**
Basic Concepts of Generation, Basic Constraints, Implication Constraints, Soft Constraints, Weighted Constraints, Order of Generation, Constraint Resolution, Do-Not-Generate Fields

6 **Procedural Flow Control**
Defining Methods, Conditional Actions, Iterative Actions, Useful Output Routines

7 **Events and Temporal Expressions**
Defining Events, Event Emission, Event Redefinition, Sampling Events, Temporal Operators, Temporal Expressions, Predefined Events

8 **Time Consuming Methods**
Defining TCMs, Invoking TCMs, Wait and Sync Actions, Gen Action, Using HDL Tasks and Functions

9 **Checking**
Packing and Unpacking, Data Checking, Temporal Checking

10 **Coverage**
Functional Coverage, Coverage Groups, Basic Coverage Item, Transition Coverage Items, Cross Coverage Items, Latency Coverage, Turning On Coverage, Viewing Coverage Results

11 **Running the Simulation**
Verification Components, Execution Flow, Synchronization between HDL Simulator and Specman Elite

CHAPTER 3

Basic *e* Concepts

In this chapter we discuss the basic constructs and conventions in *e*. These conventions and constructs are used throughout the later chapters. These conventions provide the necessary framework for understanding *e*. This chapter may seem dry, but understanding these concepts is a necessary foundation for the successive chapters.

Chapter Objectives

- Understand conventions for code segments, comments, white space, numbers, constants, and macros.

- Describe how to import other *e* files.

- Define the data types such as scalar type and subtypes, enumerated scalar type, list type, and string type.

- Understand syntax hierarchy of statements, struct members, actions, and expressions.

- Explain the use of simulator variables in *e*.

3.1 Conventions

e contains a stream of tokens. Tokens can be comments, delimiters, numbers, constants, identifiers, and keywords. *e* is a case-sensitive language. All keywords are in lowercase.

3.1.1 Code Segments

A code segment is enclosed with a begin-code marker **<'** and an end-code marker **'>**. Both the begin-code and the end-code markers must be placed at the beginning of a line (leftmost), with

no other text on that same line (no code and no comments). The example below shows three
lines of code that form a code segment.

```
<'
    import cpu_test_env;
'>
```

Several code segments can appear in one file. Each code segment consists of one or more statements.

3.1.2 Comments and White Space

e files begin as a comment which ends when the first begin-code marker <' is encountered.

Comments within code segments can be marked with double dashes (--) or double slashes (//).

```
a = 5;          -- This is an inline comment
b = 7;          // This is also an inline comment
```

The end-code '> and the begin-code <' markers can be used in the middle of code sections, to
write several consecutive lines of comment.

```
Import the basic test environment for the CPU...
This is a comment

<'
    import cpu_test_env;
'>

This particular test requires the code that bypasses bug#72 as well as
the constraints that focus on the immediate instructions. This is a
comment

<'
    import bypass_bug72;
    import cpu_test0012;
'>
```

3.1.3 Numbers

There are two types of numbers, sized and unsized.

3.1.3.1 Unsized Numbers

Unsized numbers are always positive and zero-extended unless preceded by a hyphen. Decimal
constants are treated as signed integers and have a default size of 32 bits. Binary, hex, and octal
constants are treated as unsigned integers, unless preceded by a hyphen to indicate a negative
number, and have a default size of 32 bits.

The notations shown in Table 3-1 can be used to represent unsized numbers.

Table 3-1 Representing Unsized Numbers in Expressions

Notation	Legal Characters	Examples
Decimal integer	Any combination of 0-9 possibly preceded by a hyphen - for negative numbers. An underscore (_) can be added anywhere in the number for readability.	12, 55_32, -764
Binary integer	Any combination of 0-1 preceded by *0b*. An underscore (_) can be added anywhere in the number for readability.	0b100111, 0b1100_0101
Hexadecimal integer	Any combination of 0-9 and a-f preceded by *0x*. An underscore (_) can be added anywhere in the number for readability.	0xff, 0x99_aa_bb_cc
Octal integer	Any combination of 0-7 preceded by *0o*. An underscore (_) can be added anywhere in the number for readability.	0o66_123
K (kilo: multiply by 1024)	A decimal integer followed by a K or k. For example, 32K = 32768.	32K, 32k, 128k
M (mega: multiply by 1024*1024)	A decimal integer followed by an M or m. For example, 2m = 2097152.	1m, 4m, 4M

3.1.3.2 Sized Numbers

A sized number is a notation that defines a literal with a specific size in bits. The syntax is:

width-number ' (**b|o|d|h|x**) *value-number*;

The width number is a decimal integer specifying the width of the literal in bits. The value number is the value of the literal and it can be specified in one of four radixes, as shown in Table 3-2.

> **NOTE** If the value number is more than the specified size in bits, its most significant bits are ignored. If the value number is less that the specified size, it is padded by zeros.

Table 3-2 Radix Specification Characters

Radix	Represented By	Example
Binary	A leading 'b or 'B	8'b11001010
Octal	A leading 'o or 'O	6'o45
Decimal	A leading 'd or 'D	16'd63453
Hexadecimal	A leading 'h or 'H or 'x or 'X	32'h12ffab04

3.1.4 Predefined Constants

A set of constants is predefined in *e*, as shown in Table 3-3.

Table 3-3 Predefined Constants

Constant	Description
TRUE	For boolean variables and expressions
FALSE	For boolean variables and expressions
NULL	For structs, specifies a NULL pointer; for character strings, specifies an empty string
UNDEF	UNDEF indicates NONE where an index is expected
MAX_INT	Represents the largest 32-bit **int** (2^{31} -1)
MIN_INT	Represents the largest negative 32-bit **int** (-2^{31})
MAX_UINT	Represents the largest 32-bit **uint** (2^{32}-1)

3.1.4.1 Literal String

A literal string is a sequence of zero or more ASCII characters enclosed by double quotes (" "). The special escape sequences shown in Table 3-4 are allowed.

Table 3-4 Escape Sequences in Strings

Escape Sequence	Meaning
\n	New-line
\t	Tab
\f	Form-feed
\"	Quote
\\	Backslash
\r	Carriage-return

This example shows escape sequences used in strings. Although other constructs are introduced here only for the sake of completeness, please focus only on the string syntax.

```
<'
extend sys {

    m() is {
        var header: string = //Define a string variable
          "Name\tSize in Bytes\n----\t------------\n";
        var p: packet = new;
        var pn: string = p.type().name;
        var ps: uint = p.type().size_in_bytes;
        outf("%s%s\t%d", header, pn, ps);
    };
};
'>
```

The result of running the example above is shown below.

```
Specman> sys.m()
Name    Size in Bytes
----    -------------
packet  20
```

3.1.5 Identifiers and Keywords

The following sections describe the legal syntax for identifiers and keywords.

3.1.5.1 Legal *e* Identifiers

User-defined identifiers in *e* code consist of a case-sensitive combination of any length of the characters A-Z, a-z, 0-9, and underscore. They must begin with a letter. Identifiers beginning with an underscore have a special meaning in *e* and are not recommended for general use. Identifiers beginning with a number are not allowed.

The syntax of an *e* module name (a file name) is the same as the syntax of UNIX file names, with the following exceptions.

- '@' and '~' are not allowed as the first character of a file name.
- '[', ']', '{', '}' are not allowed in file names.
- Only one '.' is allowed in a file name.

3.1.5.2 *e* Keywords

The keywords listed in Table 3-5 below are the reserved words of the *e* language. Some of the terms are keywords only when used together with other terms, such as "key" in "**list(key:**key)", "before" in "**keep gen** x **before** y", or "computed" in "**define** def **as computed**".

Table 3-5 List of Keywords

all of	all_values	and	as a	as_a
assert	assume	async	attribute	before
bit	bits	bool	break	byte
bytes	c export	case	change	check that
compute	computed	consume	continue	cover
cross	cvl call	cvl callback	cvl method	cycle
default	define	delay	detach	do
down to	dut_error	each	edges	else
emit	event	exec	expect	extend
fail	fall	file	first of	for
force	from	gen	global	hdl pathname
if	#ifdef	#ifndef	in	index
int	is	is a	is also	is c routine
is empty	is first	is inline	is instance	is not a
is not empty	is only	is undefined	item	keep
keeping	key	like	line	list of
matching	me	nand	new	nor
not	not in	now	on	only
or	others	pass	prev_	print

range	ranges	release	repeat	return
reverse	rise	routine	select	session
soft	start	state machine	step	struct
string	sync	sys	that	then
time	to	transition	true	try
type	uint	unit	until	using
var	verilog code	verilog function	verilog import	verilog simulator
verilog task	verilog time	verilog timescale	verilog trace	verilog variable
vhdl code	vhdl driver	vhdl function	vhdl procedure	vhdl driver
vhdl simulator	vhdl time	when	while	with
within				

3.1.6 Macros

The simplest way to define *e* macros is with the **define** statement. An *e* macro can be defined with or without an initial ` character.

```
<'
define WORD_WIDTH 16; //Definition of the WORD_WIDTH macro
struct t {
    f: uint (bits: WORD_WIDTH); //Usage of WORD_WIDTH macro
};
'>
```

You can also import a file with Verilog `**define** macros using the keywords **verilog import**.

```
macros.v (Verilog defines file)
`define BASIC_DELAY    2
`define TRANS_DELAY    `BASIC_DELAY+3
`define WORD_WIDTH 8


-------------------------------------------------------
dut_driver.e (e file)
<'
verilog import macros.v; //Imports all definitions from
                         //macros.v file
//Macros imported from Verilog must be used
//with a preceding `.
struct dut_driver {
    ld: list of int(bits: `WORD_WIDTH); //use verilog macro
    keep ld.size() in [1..`TRANS_DELAY];//use verilog macro
};
'>
```

3.1.7 Importing *e* Files

e files are called modules. An *e* file can import another *e* file using the **import** keyword. The **import** statement loads additional *e* modules before continuing to load the current file. If no extension is given for the imported file name, a ".e" extension is assumed. The modules are loaded in the order they are imported. The **import** statements must be before any other statements in the file.

```
//File Name: pci_transaction_definition.e
<'
type PCICommandType: [ IO_READ=0x2, IO_WRITE=0x3,
                       MEM_READ=0x6, MEM_WRITE=0x7 ];
struct pci_transaction {
    address: uint;
    command: PCICommandType;
    bus_id: uint;
};
'>
//End File: pci_transaction_definition.e
-----------------------------------------------------------
//File Name: pci_transaction_extension.e
<'
//Import the file defined above. Note that the .e
//extension is assumed in an import statement
import pci_transaction_definition; //.e extension is the default

extend pci_transaction {
    data: list of uint;
};
'>
//End File: pci_transaction_extension.e
```

If a specified module has already been loaded or compiled, the statement is ignored. For modules not already loaded or compiled, the search sequence is:

  1. The current directory

  2. Directories specified by the SPECMAN_PATH[1] environment variable

  3. The directory in which the importing module resides

3.2 Data Types

This section discusses the basic data types in *e*.

---

1.   This is an environment variable used by Specman Elite for setting up search paths.

3.2.1 Scalar Types

Scalar types in *e* are one of the following:

- Numeric

- Boolean

- Enumerated

3.2.1.1 Numeric and Boolean Scalar Types

Table 3-6 shows predefined numeric and boolean types in *e*.

Table 3-6 Scalar Types

Type Name	Function	Usage Example
int	Represents numeric data, both negative and non-negative integers. Default Size = 32 bits	length: int; addr: int (bits:24); // 24-bit vector
uint	Represents unsigned numeric data, non-negative integers only. Default Size = 32 bits	delay: uint; addr: uint (bits:16); // 8-bit vector
bit	An unsigned integer in the range 0–1. Size = 1 bit	valid: bit; // 1-bit field

byte	An unsigned integer in the range 0–255.
	Size = 8 bits
		data: byte; // 8-bit field
		data: uint (bits:8); // Equivalent
		//definition

time	An integer in the range $0-2^{63}-1$. Default Size = 64 bits	delay: time; //64-bit time variable

bool	Represents truth (logical) values, TRUE (1) and FALSE (0). Default Size = 1 bit	frame_valid: bool; //TRUE or
		//FALSE

3.2.1.2 Enumerated Scalar Types

Enumerated types define the valid values for a variable or field as a list of symbolic constants. For example, the following declaration defines the variable *instr_kind* as having two legal values.

```
<'
//Implicit enumerated type. immediate=0, register=1
type instr_kind: [immediate, register];
'>
```

These symbolic constants have associated unsigned integer values. By default, the first name in the list is assigned the value zero. Subsequent names are assigned values based upon the maximum value of the previously defined enumerated items + 1.

It is also possible to assign explicit unsigned integer values to the symbolic constants. This method is used when the enumerated types may not be defined in a particular order.

```
<'
//Explicit enumerated type. immediate=4, register=8
type instr_kind: [immediate=4, register=8];
'>
```

It is sometimes convenient to introduce a named enumerated type as an empty type.

```
<'
type packet_protocol: []; //Define empty type
'>
```

Once the protocols that are meaningful in the program are identified the definition of the type can be extended.

```
<'
//Extend this type in a separate file.
//No need to touch the original file.
extend packet_protocol : [Ethernet, IEEE, foreign];

//Define a struct that uses a field of the above type.
struct packet {
kind: packet_protocol; //Define a field of type
                       //packet_protocol. Three possible values.
};
'>
```

3.2.1.3 Scalar Subtypes

A subtype can be created from one of the following:

- A predefined numeric or boolean type (**int**, **uint**, **bool**, **bit**, **byte**, **time**)
- A previously defined enumerated type
- A previously defined scalar subtype

Creation of subtypes is shown in the example below.

```
<'
//Define an enumerated type opcode
type opcode: [ADD, SUB, OR, AND];

//Define a subtype using the previously defined type
//This subtype includes opcodes for logical operations
//OR, AND
type logical_opcode: opcode [OR, AND];

//Define a subtype of a predefined scalar type, 4 bit
//unsigned vector
type small: uint(bits:4);

//Define a struct that uses the above types
struct instruction {
op1: opcode; //Field of type opcode
op2: logical_opcode; //Field of type logical_opcode
length: small; //4 bit unsigned vector
};

'>
```

3.2.2 List Types

List types hold ordered collections of data elements. Items in a list can be indexed with the subscript operator [ ], by placement of a non-negative integer expression in the brackets. List indexes start at zero. You can select an item from a list by specifying its index. For example, my_list[0] refers to the first item in the list named my_list. Lists can be of any type. However, a list of lists is not allowed. All items in a list must be of the same type. Lists are dynamically resizable and they come with many predefined methods.

Lists are defined by using the **list of** keyword in a variable or a field definition.

```
<'
struct packet {
addr: uint(bits:8); // 8-bit vector
data1: list of byte; //List of 8-bit values
};
```

```
(continued...)
struct sys {
packets[10]: list of packet; //List of 10 packet structures
values: list of uint(bits:128); //List of 128-bit vectors
};
'>
```

e does not support multidimensional lists (lists of lists). To create a list with sublists in it, you can create a **struct** to contain the sublists, and then create a list of such structs as the main list. In the example above, the packet struct contains a list of bytes. In **sys** struct, there is a list of 10 packets. Thus, **sys** contains a list of lists.

3.2.2.1 Keyed Lists

A keyed list data type is similar to hash tables or association lists found in other programming languages. Keyed lists are defined with the keyword **key**. The declaration below specifies that *packets* is a list of packets, and that the *protocol* field in the *packet* type is used as the hash key. Keyed lists are very useful for searching through a list with a key.

```
<'
type packet_protocol : [Ethernet, IEEE, foreign];
struct packet {
    protocol: packet_protocol;
};
struct dtypes {
    m() is { //Method definition explained later in book
        // Define local variable, keyed list
        var packets : list (key: protocol) of packet;
    };
};
'>
```

3.2.3 The String Type

A string data type contains a series of ASCII characters enclosed by quotes (" "). An example of string declaration and initialization is shown below.

```
<'
struct dtypes {
    m() is { //Define a method (procedure)
      var message: string; //Define a variable of type string
      message = "Beginning initialization sequence…";
                          //String value
      print message; //Print string
    };
};
'>
```

3.3 Simulator Variables

In Chapter 2, we discussed two hierarchies in an *e*-based environment.

1. The design hierarchy represented in Verilog or VHDL.

2. The verification hierarchy represented in *e*

An example of the two hierarchies is shown in Figure 3-1 below

Figure 3-1 Verification and Design Hierarchies

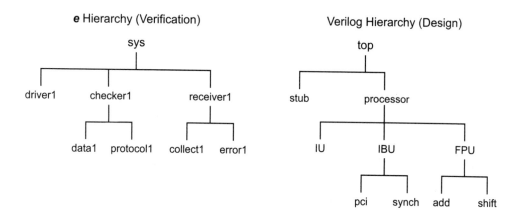

Any *e* structure should be able to access the simulator variables for reading or writing during simulation. The subsections below explain how to read and write simulator variables.

3.3.1 Driving and Sampling DUT Signals

In *e*, one can access simulator variables by simply providing the hierarchical path to the variable within single quotes (''). The example below shows how to access the design hierarchy shown in Figure 3-1.

```
<'
struct driver{ //Struct in the e environment
r_value: uint(bits:4); //Define a 4 bit field to read

read_value() is { //Define a method to read simulator
                  //variable
   //Right hand side is the simulator variable
   //operand is a variable in module add in Verilog/VHDL
   //The "/" represents the traversal of hierarchy
   //Left hand side is an e field in struct driver
   r_value = '~/top/processor/FPU/add/operand';
};

write_value() is { //Define a method to write simulator
                   //variable
   //Left hand side is the simulator variable
   //Right hand side can be a constant, an e variable,
   //or a simulator variable.
   '~/top/processor/FPU/add/operand' = 7; //Write 7 to variable
};
'>
```

3.3.2 Computed Signal Names

While accessing simulator variables, one can compute all or part of the signal name at run time by using the current value of an *e* variable inside parentheses. For example, in Figure 3-1, if there are three processors, *processor_0*, *processor_1*, and *processor_2* instantiated inside top, it should be possible to pick out one of the three processor instances based on an *e* variable.

e allows the usage of many parentheses in the computation of a signal name. Thus, it is possible to dynamically choose the signal name at run time based on *e* variable values.

```
<'
struct driver{ //Struct in the e environment
id: uint(bits:2); //2 bit ID field determines processor 0,1,2
r_value: uint(bits:4); //Define a 4 bit field to read

read_value() is { //Define a method to read simulator
                  //variable
   r_value = '~/top/processor_(id)/FPU/add/operand';

   //If the id field == 1, then the above assignment
   //will be dynamically set to
   //r_value = '~/top/processor_1/FPU/add/operand';
};

'>
```

3.4 Syntax Hierarchy

Unlike Verilog or C, *e* enforces a very strict syntax hierarchy. This is very useful when one is writing or loading *e* code. Based on the error messages during loading, it is easy to determine the nature of the syntactical mistake. The strict hierarchy also makes it very difficult to make mistakes.

Every *e* construct belongs to a construct category that determines how the construct can be used. The four categories of *e* constructs are shown in Table 3-7 below.

Table 3-7 Constructs in the Syntax Hierarchy

Statements	Statements are top-level constructs and are valid within the begin-code <' and end-code '> markers.
Struct members	Struct members are second-level constructs and are valid only within a struct definition.
Actions	Actions are third-level constructs and are valid only when associated with a struct member, such as a method or an event.
Expressions	Expressions are lower-level constructs that can be used only within another *e* construct.

Figure 3-2 shows an example of the strict syntax hierarchy.

Figure 3-2 Syntax Hierarchy

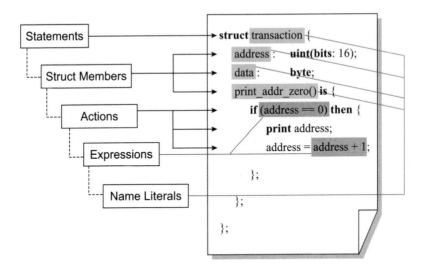

The following sections describe each element of the syntax hierarchy in greater detail. Henceforth, any syntactical element in the book will be described as a statement, struct member, action, or expression.

3.4.1 Statements

Statements are the *top-level* syntactic constructs of the *e* language and perform the functions related to extending the *e* language and interfacing with the simulator.

Statements are valid within the begin-code <' and end-code '> markers. They can extend over several lines and are separated by semicolons. For example, the following code segment has two statements.

```
<'
    import bypass_bug72; //Statement to import bypass_bug72.e
    type opcode: [ADD, SUB]; //Statement to define
                             //enumerated type
'>
```

Table 3-8 shows the complete list of *e* statements.

Table 3-8 *e* Statements

struct	Defines a new data structure.
unit	Defines a new data structure with special properties.
type	Defines an enumerated data type or scalar subtype.
extend	Modifies a previously defined struct or type.
define	Extends the *e* language by defining new commands, actions, expressions, or any other syntactic element.
#ifdef, #ifndef	Is used with **define** statements to place conditions on the *e* parser.
routine ... is C routine	Declares a user-defined C routine that you want to call from *e*.
C export	Exports an *e* declared type or method to C.
import	Reads in an *e* file.
verilog import	Reads in Verilog macro definitions from a file.
verilog code	Writes Verilog code to the stubs file, which is used to interface *e* programs with a Verilog simulator.
verilog time	Specifies Verilog simulator time resolution.
verilog variable reg | wire	Specifies a Verilog register or wire that you want to drive from *e*.
verilog variable memory	Specifies a Verilog memory that you want to access from *e*.
verilog function	Specifies a Verilog function that you want to call from *e*.
verilog task	Specifies a Verilog task that you want to call from *e*.

Table 3-8 *e* Statements (Continued)

vhdl code	Writes VHDL code to the stubs file, which is used to interface *e* programs with a VHDL simulator.
vhdl driver	Is used to drive a VHDL signal continuously via the resolution function.
vhdl function	Declares a VHDL function defined in a VHDL package.
vhdl procedure	Declares a VHDL procedure defined in a VHDL package.
vhdl time	Specifies VHDL simulator time resolution.

3.4.2 Struct Members

Struct member declarations are second-level syntactic constructs of the *e* language that associate the entities of various kinds with the enclosing **struct** or **unit**.

Struct members can only appear inside a **struct** definition statement. They can extend over several lines and are separated by semicolons. For example, the following struct "packet" has three struct members, *len*, *data*, and a method *m( )*.

```
<'
struct packet{
    len: int; //Field of a struct
    data[len]: list of byte; //Field of a struct
    m() is { //Method(procedure) is struct member
    --

    --
    };
};
'>
```

A struct can contain multiple struct members of any type in any order. Table 3-9 shows a brief description of *e* struct members. This list is not comprehensive. See Appendix A for a description of all struct members.

Table 3-9 *e* Struct Members

Field declaration	Defines a data entity that is a member of the enclosing struct and has an explicit data type.
Method declaration	Defines an operational procedure that can manipulate the fields of the enclosing struct and access run time values in the DUT.
Subtype declaration	Defines an instance of the parent struct in which specific struct members have particular values or behavior.
Constraint declaration	Influences the distribution of values generated for data entities and the order in which values are generated.
Coverage declaration	Defines functional test goals and collects data on how well the testing is meeting those goals.
Temporal declaration	Defines *e* events and their associated actions.

3.4.3 Actions

e actions are lower-level procedural constructs that can be used in combination to manipulate the fields of a struct or exchange data with the DUT. Actions are associated with a struct member, specifically a method, an event, or an "on" struct member. Actions can also be issued interactively as commands at the command line.

Actions can extend over several lines and are separated by semicolons. An action block is a list of actions separated by semicolons and enclosed in curly braces, { }.

Shown below is an example of an action (an invocation of a method, "transmit()") associated with an event, *xmit_ready*. Another action, **out()** is called in the *transmit()* method.

```
<'
struct packet{
    //Declare an event (struct member)
    event xmit_ready is rise('~/top/ready');
    //Declare fields (struct members)
    length: byte;
    delay: uint;
    //Declare an on construct (a struct member)
    //The transmit method is called in the on construct
    //See "On Struct Member" on page 184 for details.
    on xmit_ready {
        transmit(); //Call transmit method (Action)
    };
    //Declare a method (a struct member)
    transmit() is {
      length = 5; //Action that sets value of length
      delay = 10; //Action that sets value of delay
      out("transmitting packet..."); //Action to print
                                     //message
    };
};
'>
```

The following sections describe the various categories of *e* actions. These actions can be used only inside method declarations or the **on** struct member. Details on the usage of these actions are not provided in these sections but will be treated in later chapters.

3.4.3.1 Actions for Creating or Modifying Variables

The actions described in Table 3-10 are used to create or modify *e* variables.

Table 3-10 Actions for Creating or Modifying Variables

var	Defines a local variable.
=	Assigns or samples values of fields, local variables, or HDL objects.
op	Performs a complex assignment (such as add and assign, or shift and assign) of a field, local variable, or HDL object.

Table 3-10 Actions for Creating or Modifying Variables (Continued)

force	Forces a Verilog net or wire to a specified value, overriding the value driven from the DUT (rarely used).
release	Releases the Verilog net or wire that was previously forced.

3.4.3.2 Executing Actions Conditionally

The actions described in Table 3-11 allow conditional behavior to be specified in *e*.

Table 3-11 Executing Actions Conditionally

if then else	Executes an action block if a condition is met and a different action block if it is not.
case *labeled-case-item*	Executes one action block out of multiple action blocks depending on the value of a single expression.
case *bool-case-item*	Evaluates a list of boolean expressions and executes the action block associated with the first expression that is true.

3.4.3.3 Executing Actions Iteratively

The actions described in Table 3-12 implement looping in *e*.

Table 3-12 Executing Actions Iteratively

while	Executes an action block repeatedly until a boolean expression becomes FALSE.
repeat until	Executes an action block repeatedly until a boolean expression becomes TRUE.
for each in	For each item in a list that is a specified type, executes an action block.
for from to	Executes an action block for a specified number of times.
for	Executes an action block for a specified number of times.

Table 3-12 Executing Actions Iteratively (Continued)

for each line in file	Executes an action block for each line in a file.
for each file matching	Executes an action block for each file in the search path.

3.4.3.4 Actions for Controlling Loop Execution

The actions described in Table 3-13 control the execution of loops.

Table 3-13 Actions for Controlling Loop Execution

break	Breaks the execution of the enclosing loop.
continue	Stops execution of the enclosing loop and continues with the next iteration of the same loop.

3.4.3.5 Actions for Invoking Methods and Routines

The actions described in Table 3-14 illustrate the ways for invoking methods (*e* procedures) and routines (C procedures).

Table 3-14 Actions for Invoking Methods and Routines

method()	Calls a regular method.
tcm()	Calls a TCM.
start *tcm()*	Launches a TCM as a new thread (a parallel process).
routine()	Calls an *e* predefined routine.
Calling C routines from e	Describes how to call user-defined C routines.
compute_method()	Calls a value-returning method without using the value returned.
return	Returns immediately from the current method to the method that called it.

3.4.3.6 Time Consuming Actions

The actions described in Table 3-15 may cause simulation time to elapse before a callback is issued by the Simulator to Specman Elite.

Table 3-15 Time Consuming Actions

emit	Causes a named event to occur.
sync	Suspends execution of the current TCM until the temporal expression succeeds.
wait	Suspends execution of the current TCM until a given temporal expression succeeds.
all of	Executes multiple action blocks concurrently, as separate branches of a fork. The action following the **all of** action is reached only when all branches of the **all of** have been fully executed.
first of	Executes multiple action blocks concurrently, as separate branches of a fork. The action following the **first of** action is reached when any of the branches in the **first of** has been fully executed.
state machine	Defines a state machine.

3.4.3.7 Generating Data Items

The action described in Table 3-16 is useful for generating the fields in data items based on spec-

Table 3-16 Generating Data Items

gen	Generates a value for an item, while considering all relevant constraints.

ified constraints.

3.4.3.8 Detecting and Handling Errors

The actions described in Table 3-17 are used for detecting and handling errors.

3.4.3.9 General Actions

The actions described in Table 3-18 are used for printing and setting configuration options for various categories.

Table 3-17 Detecting and Handling Errors

check that	Checks the DUT for correct data values.
expect	Expects a certain temporal expression to succeed.
dut_error()	Defines a DUT error message string.
assert	Issues an error message if a specified boolean expression is not true.
warning()	Issues a warning message.
error()	Issues an error message when a user error is detected.
fatal()	Issues an error message, halts all activities, and exits immediately.
try()	Catches errors and exceptions.

Table 3-18 General Actions

print	Prints a list of expressions.
set_config()	Sets options for various categories, including printing.

3.4.4 Expressions

Expressions are constructs that combine operands and operators to represent a value. The resulting value is a function of the values of the operands and the semantic meaning of the operators. A few examples of operands are shown below.

```
address + 1
a + b
address == 0
'~/top/port_id' + 1
```

Expressions are combined to form actions. Each expression must contain at least one operand, which can be:

- A literal value (an identifier)
- A constant
- An *e* entity, such as a method, field, list, or struct
- An HDL entity, such as a signal

A compound expression applies one or more operators to one or more operands. Strict type checking is enforced in *e*.

3.4.5 Name Literals (Identifiers)

Identifiers are names assigned to variables, fields, structs, units, etc. Thus identifiers are used at all levels of the syntax hierarchy, i.e., they are used in statements, struct members, actions, and expressions. Identifiers must follow the rules set in "Legal *e* Identifiers" on page 40. In the example below, identifiers are highlighted with comments.

```
<'
struct packet{ //"packet" is an identifier
    %address: uint(bits:2); // "address" is an identifier
    %len: uint(bits:6); //"len" is an identifier
    %data[len]: list of byte; //"data" is an identifier
    my_method() is { //"my_method" is an identifier
    result = address + len; //Identifiers are used in
                            //an expression

    --
    };
};
'>
```

3.5 Summary

We discussed the basic concepts of *e* in this chapter. These concepts lay the foundation for the material discussed in the further chapters.

- Conventions for code segments, comments, white space, numbers, constants, and macros were discussed.

- It is possible to import other *e* files in an *e* file. The **import** statements must always be the first statements in the file.

- *e* contains data types such as scalar type and subtypes, enumerated scalar type, list type, and string type.

- Simulator variables can be directly read or written from *e* code. One can access simulator variables by enclosing the hierarchical path to the simulator variable in single quotes.

- There is a strict syntax hierarchy of statements, struct members, actions, and expressions. A strict hierarchy enforces coding discipline and minimizes errors.

3.6 Exercises

1. Write a code segment that contains just one statement to import file *basic_types.e*.

2. Determine which comments in the following piece of *e* code are written correctly. Circle the comments that are written incorrectly.

```
<'
Import the basic test environment for the CPU...
    import cpu_test_env;
'>

This particular test requires the code that bypasses bug#72 as well as
the constraints that focus on the immediate instructions.

<'
    /*Import the bypass bug file*/
    import bypass_bug72;
    //Import the cpu test file_0012
    import cpu_test0012;
    --Import the cpu test file_0013
    import cpu_test0013;

'>
```

3. Practice writing the following numbers. Use _ for readability.

   a. Decimal number 123 as a sized 8-bit number in binary
   b. A 16-bit hexadecimal with decimal value 135
   c. An unsized hex number 1234
   d. 64K in decimal
   e. 64K in *e*

4. Name the predefined constants in *e*.

5. Write the correct string to produce each of the following outputs.

   a. "This is a string displaying the % sign"
   b. "out = in1 + in2 \\"
   c. "Please ring a bell \t"
   d. "This is a backslash \ character\n"

6. Determine whether the following identifiers are legal:

    a. system1
    b. 1reg
    c. ~latch
    d. @latch
    e. exec[
    f. exec@

7. Define a macro *LENGTH* equal to 16.

8. Declare the following fields in *e*.

    a. An 8-bit unsigned vector called *a_in*
    b. A 24-bit unsigned vector called *b_in*
    c. A 24-bit signed vector called *c_in*
    d. An integer called *count*
    e. A time field called *snap_shot*
    f. A string called *l_str*
    g. A 1-bit field called *b*

9. Declare an enumerated type called *frame_type*. It can have values SMALL and LARGE. Declare a field called *frame* of this enumerated type.

10. Write the following actions with simulator variables.

    a. Assign the value of simulator variable *top.x1.value* to *len* field in *e*.
    b. Write the value 7 to the simulator variable *top.x1.value*.

11. Describe the four levels of *e* syntax hierarchy. Identify three keywords used at each level of the hierarchy.

Creating Hierarchy with Structs and Units

In the previous chapters, we acquired an understanding of the fundamental *e* modeling concepts, basic conventions, and *e* constructs. In this chapter, we take a closer look at building hierarchy with *e* using structs and units.

Chapter Objectives

- Describe the definition of a **struct**.
- Explain how to extend an existing **struct** definition.
- Understand how to define fields.
- Describe the use of list fields.
- Understand how to create **struct** subtypes for variations of the basic **struct** definition.
- Describe how to extend **struct** subtypes.
- Explain the advantages of units.
- Explain differences between structs and units.
- Understand how to define a **unit**.

4.1 Defining Structs

Structs are used to define data elements and the behavior of components in a verification environment. A struct can hold all types of data and methods. Structs are defined with the keyword

struct. Struct definitions are statements and are at the highest level in the *e* hierarchy as discussed in "Syntax Hierarchy" on page 49. The syntax definition of a **struct** is as follows:

```
struct struct-type [like base-struct-type] {
        [struct-member] [;struct-member] …};
```

The components defined in the above **struct** definition are explained in detail in Table 4-1 below.

Table 4-1 Components of a Struct Definition

Name	Description
struct-type	The name of the new struct type
base-struct-type	The type of the struct from which the new struct inherits its members
struct-member; …	The contents of the struct; the following are types of struct members: • data fields for storing data • methods for procedures • events for defining temporal triggers • coverage groups for defining coverage points • **when**, for specifying inheritance subtypes • declarative constraints for describing relations between data fields • **on**, for specifying actions to perform upon event occurrences • **expect**, for specifying temporal behavior rules • the definition of a struct can be empty, containing no members

Example 4-1 shows a simple **struct** definition containing only fields and methods. This definition does not contain all struct members described in Table 4-1. Some of these struct members will be explained later in this book.

Example 4-1 Basic Struct Definition

```
<'
type packet_kind: [atm, eth]; // Enumerated type

//Definition packet struct
struct packet {
    len: int; //Field of struct
    keep len < 256; //Constraint on struct
    kind: packet_kind; //Field of struct
    calc_par() is { //Method (procedure) in a struct
    --
    --
    }; //end of method definition
}; //end of packet struct
'>
```

4.2 Extending Structs

Struct extensions add struct members to a previously defined **struct**. These extensions can be specified in the same file or a different file. Moreover, it is possible to apply many extensions to a **struct**. Each extension adds a layer of struct members to the original struct definition. The syntax definition for extending a **struct** is as follows:

```
extend [struct-subtype] base-struct-type {
          [struct-member] [;struct-member] ...};
```

An extension to a struct can have all components that are contained in a struct definition. The components that can be defined in an extension of the **struct** are explained in further detail in Table 4-2.

Table 4-2 Components of a Struct Extension

base-struct-type	Indicates the base struct type to extend.
member; ...	Denotes the contents of the struct. Any legal struct member is allowed in a struct extension.
struct-subtype	Adds struct members to the specified subtype of the base struct type only. The added struct members are known only in that subtype, not in other subtypes. This will be discussed later.

Example 4-2 shows the extension of a **struct** definition in the same file. Note that **sys** is a predefined struct in *e*. Therefore, **sys** can only be extended. All user-defined structs must be instantiated below **sys**.

Example 4-2 Extension of a Struct Definition in the Same File

```
File name: packet.e
-------------------
<'
type packet_kind: [ATM, ETH]; // Enumerated type

//Definition packet struct
struct packet {
    len: int; //Field of struct
    keep len < 256; //Constraint on struct
    kind: packet_kind; //Field of struct
    calc_par() is { //Method (procedure) in a struct
    --
    }; //end of method definition
}; //end of packet struct
```

Example 4-2 Extension of a Struct Definition in the Same File (Continued)

```
<'
//Extension in the same file
//Extend the sys struct and instantiate a list
//of packets in sys. Initially, sys is empty.
//Extensions add struct members and instantiations
//to sys
extend sys {
    packets: list of packet;
    keep packets.size() == 10;
    run() is also {
        print packets;
    };
};

//Extension of packet definition in the
//same file
extend packet {
    addr : byte; //Add 8-bit address field to definition
    keep len < 128; //Extend original packet definition
};

'>
```

A struct can also be extended in a separate file as shown in Example 4-3. The original struct definition must be loaded before the extension file is loaded.

Example 4-3 Extension of a Struct Definition in a Different File

```
File name: packet_extension.e
-----------------------------
<'
//Import the original packet definition
import packet.e;

//Extension of packet struct in a separate file
extend packet {
    keep kind == ATM; //Add constraint to packet definition
                      //to keep all packets of type ATM
    keep len == 64;   //Add constraint to packet definition
                      //to keep length exactly 64.
};

'>
```

The advantage of using the **extend** statement is that the original definition of the struct does not need to be modified or edited. This makes it very easy to quickly create a basic verification environment and then enhance it incrementally. Figure 4-1 shows a representation of how an **extend** statement adds code incrementally to the definition of the *packet* struct. Specman Elite gathers the original struct definition and all extensions to the definition to create a description of each struct.

Figure 4-1 Extending the Packet Struct

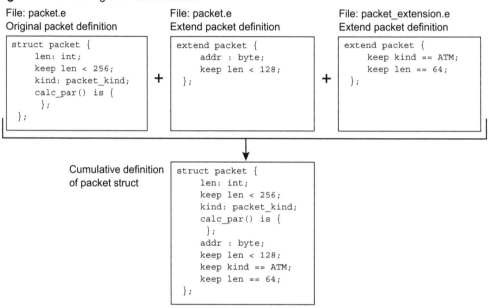

The following sections discuss the fields of a struct definition in greater detail.

4.3 Defining Fields

Each struct definition contains a number of fields. These fields together create a component of the verification environment. The values of these fields can be generated by Specman Elite based on the constraints provided for each field. The general syntax for defining fields is as follows.

```
[!][%] field-name[: type] [[min-val .. max-val]] [((bits|bytes):num)];
```

The components of a field definition are described in Table 4-3.

Unless you define a field as ungenerated, Specman Elite will generate a value for it when the struct is generated, subject to any constraints that exist for the field. However, even for generated

Table 4-3 Components of a Field Definition

!	Denotes an ungenerated field. The value for this field is not assigned during the generation phase in Specman Elite. The generation mechanism is explained in detail in later chapters. This is useful for fields that are to be explicitly assigned during the simulation, or whose values involve computations that cannot be expressed in constraints.
	Ungenerated fields get default initial values (0 for scalars, NULL for structs, empty list for lists). An ungenerated field whose value is a range (such as [0..100]) gets the first value in the range. If the field is a struct, it will not be allocated and none of the fields in it will be generated.
	The "!" and "%" options can be used together, in either order.
%	Denotes a physical field. If a field is not prefixed with "%", it means that the field is a virtual field. Fields that represent data that are to be sent to the HDL device in the simulator or that are to be used for memories in the simulator or in Specman Elite need to be physical fields. The importance of physical fields is explained in detail in later chapters.
	The "!" and "%" options can be used together, in either order.
field-name	Indicates the name of the field being defined.
type	Indicates the type for the field. This can be any scalar type, string, struct, or list.
	If the field name is the same as an existing type, you can omit the ": *type*" part of the field definition. Otherwise, the type specification is required. However, omitting the ": *type*" part of the field definition is not a recommended practice.
min-val..max-val	Indicates an optional range of values for the field, in the form [0..10]. If no range is specified, the range is the default range for the field's type.
(**bits** | **bytes:** *num*)	Indicates the width of the field in bits or bytes. This syntax allows you to specify a width for the field other than the default width.
	This syntax can be used for any scalar field, even if the field has a type with a known width.

fields, you can always assign values in user-defined methods or predefined methods. The ability to assign a value to a field is not affected by either the "!" option or generation constraints.

Example 4-4 shows a struct definition that contains several types of fields.

Example 4-4 Field Definitions in a Struct

```
Example of different types of field definitions in a struct
<'
//Define an enumerated type with explicitly assigned
//values to each enumeration. This enumerated type is
//defined to be 16 bits wide.
type NetworkType: [IP=0x0800, ARP=0x8060] (bits: 16);

//Struct header has two physical fields denoted by % sign.
struct header {
    %address: uint (bits: 48); //48-bit unsigned vector physical field.
    %length: uint [0 .. 32]; //Unsigned integer physical field.
};

//Struct packet
struct packet {
    hdr_type: NetworkType; //An enumerated field of type IP or ARP.
                           //This is a virtual field.
    %hdr: header; //A physical field "hdr" that holds an instance
                  //of the "header" struct
    is_legal: bool; //Boolean virtual field.
    !counter: uint; //Counter field that is not generated when
                    //the packet instance is generated. The ! inhibits
                    //the generation of a field.
};

//Extend the sys struct and instantiate a packet struct
extend sys {
    packet_i: packet; //Define a field packet_i
                      //of type packet. Since packet
                      //is a struct, such field definition is called
                      //instantiation.
};
'>
```

The following section discusses list fields.

4.4 Defining List Fields

List fields are used very frequently in *e*. Lists of any data type, enumerated type, or user-defined struct can be defined in *e*.

An initial size can be specified for the list. The list initially contains that number of items. The size conforms to the initialization rules, the generation rules, and the packing rules. Even if an initial size is specified, the list size can change during the test if the list is operated on by a list method that changes the number of items. All list items are initialized to their default values when the list is created. For a generated list, the initial default values are replaced by generated values. The syntax for a list declaration is shown below.

```
[!] [%] list-name[[length-exp]]: list of type ;
```

The components of a list field definition are shown in Table 4-4 below.

Table 4-4 Components of a List Field Definition

!	Do not generate this list. The "!" and "%" options can be used together, in either order.
%	Denotes a physical list. The "!" and "%" options can be used together, in either order.
list-name	Names the list being defined.
length-exp	Indicates an expression that gives the initial size for the list. The expression must evaluate to a non-negative integer.
type	Indicates he type of items in the list. This can be any scalar type, string, or struct. It cannot be a list.

Lists are dynamic objects that are used for holding stimulus items, struct instances, data values, etc. Lists can be resized. Example 4-5 illustrates a list definition.

Example 4-5 List Field Definition

```
Example of a list field definition
<'
//Define a struct cell
struct cell {
    %data: list of byte; //Physical Field: List of 8-bit values
    %length: uint; //Physical Field: Unsigned integer
    strings: list of string; //Virtual Field: List of Strings
};
//Define a struct packet
```

Example 4-5 List Field Definition (Continued)

```
struct packet {
    %is_legal: bool; //Physical Field: Boolean
    cells: list of cell; //Virtual Field: List of struct cell
                          //This field will contain multiple instances
                          //of struct cell
};
//Extend the sys struct
extend sys {
    packets[16]: list of packet; //Virtual Field: List of 16 instances
                                  //of packet struct
};
'>
```

4.4.1 List Operations

Many pseudo-methods are available to operate on lists. When a list field is defined, the pseudo-methods can be used with that list field. Table 4-5 shows the commonly used pseudo-methods that are available to operate on a list. Any arguments required by the predefined method go in parentheses after the predefined method name. Pseudo-methods can be called in actions or constraints.

Once a list field or variable has been declared, you can operate on it with a list predefined method by attaching the predefined method name, preceded by a period, to the list name. Many of the list pseudo-methods take expressions as arguments and operate on every item in the list. Table 4-5 shows the most commonly used pseudo-methods available for lists.

Table 4-5 Pseudo-methods for Lists

Name	Arguments	Description	Example
add(*item*); **push**(*item*); **add**(*list*);	Item of the same list type OR List of same type.	Adds the item or list to the end of the list.	var i_list: list of int; i_list.add(5);
add0(*item*); **push0**(*item*); **add0**(*list*);	Item of the same list type OR List of same type.	Adds the item or list to the start of the list.	var i_list: list of int; i_list.add0(5);
clear();	None	Deletes all items from the list.	var a_list; a_list.clear();

Table 4-5 Pseudo-methods for Lists (Continued)

Name	Arguments	Description	Example

delete(*index*);	Non-negative integer index.	Deletes the item at index location. Index=0 is the first item in the list.	var l_list: list of int = {2; 4; 6; 8};
			l_list.delete(2);
			//Deletes 6 from list
			//Result {2; 4; 8};

insert(*index,item*);
insert(*index, list*);
	Non-negative integer index.
	Item of the same list type OR
	List of same type.
		Inserts the item or list at the specified index.	var l_list := {10; 20; 30};
			l_list.insert(1, 99);
			//Result {10; 99; 20; 30};

pop(*item*);	Item of the same list type.	Removes the last item from the list and returns it.	var l_list := {10; 20; 30};
			l_list.pop();
			//Result {10; 20};
			//Return value = 30

pop0(*item*);	Item of the same list type.	Removes the first item from the list and returns it.
		Same as delete(0).
			var l_list := {10; 20; 30};
			l_list.pop0();
			//Result {20; 30};
			//Return value = 10

resize(*size*);	Integer indicating size to which the list is resized.	Resizes the list to the declared size and initializes all elements to the default value unless specified otherwise.	extend sys {
			run() is also {
			var q_list: list of byte;
			q_list.resize(200);
			//Resizes list to 200 bytes
			};
			};

Table 4-5 Pseudo-methods for Lists (Continued)

Name	Arguments	Description	Example

size()	None	Returns an integer equal to the number of items in the list. A common use for this method is in a **keep** constraint, to specify an exact size or a range of values for the list size.	<'
			 extend sys {
			 p_list: list of packet;
			 keep p_list.size() == 10;
			 //Size of generated list is
			 //exactly 10.
			 };
			 '>

is_empty()	None	Checks if the list is empty and returns a boolean value. Returns TRUE if list is empty, or FALSE if the list is not empty.	<'
			 extend sys {
			 b_list[5]: list of byte;
			 lmeth() is {
			 var no_l: bool;
			 no_l =
			     sys.b_list.is_empty();
			 //no_l will be FALSE
			 };
			 };
			 '>

4.4.2 Keyed Lists

Keyed lists are used to enable faster searching of lists by designating a particular field or value which is to be searched for. A keyed list can be used, for example, in the following ways:

- As a hash table, in which searching only for a key avoids the overhead of reading the entire contents of each item in the list.
- For a list that has the capacity to hold many items, but which in fact contains only a small percentage of its capacity, randomly spread across the range of possible items. An example is a sparse memory implementation.

Although all of the operations that can be done using a keyed list can also be done using a regular list, using a keyed list provides an advantage in the greater speed of searching a keyed list. The general syntax for defining a keyed list is as follows.

```
! [%] list-name: list(key: key-field) of type ;
```

The components of a keyed list definition are described in Table 4-6 below.

Table 4-6 Defining Keyed Lists

!	Do not generate this list. For a keyed list, the "!" is required, not optional.
%	Denotes a physical list. The "%" option may precede or follow the "!".
list-name	Names the list being defined.
key-field	Indicates the key of the list. For a list of structs, it is the name of a field of the struct. This is the field or value which the keyed list pseudo-methods will check when they operate on the list. For a list of scalars, the key can be the **it** variable referring to each item. The keyword **it** is context dependent and points to the relevant object in a particular context. See Table 4-7 and Example 4-6 for details on usage of the **it** variable.
type	Indicates the type of items in the list. This can be any scalar type, string, or struct. It cannot be a list.

Besides the **key** parameter, the keyed list syntax differs from regular list syntax in the following ways:

- The list must be declared with the "!" do-not-generate operator. This means that you must build a keyed list item by item, since you cannot generate it.

- The "[*exp*]" list size initialization syntax is not allowed for keyed lists. That is, "*list*[*exp*]: **list(key:** *key*) **of** *type*" is not legal. Similarly, you cannot use a **keep** constraint to constrain the size of a keyed list.

A keyed list is a distinct type, different from a regular list. This means that you cannot assign a keyed list to a regular list, nor assign a regular list to a keyed list. Table 4-7 shows three pseudo-methods associated with keyed lists.

Table 4-7 Keyed List Pseudo-methods

Name	Arguments	Description	Example
key(*key-exp*);	The key of the item that is to be returned	Returns the list item that has the specified key, or NULL if no item with that key exists in the list.	var l_list: list(key: it) of int = {5; 4; 3; 2; 1}; print l_list.key(5); //prints 5
key_index (*key-exp*);	The key to be searched for	Returns the integer index of the item that has the specified key, or returns UNDEF if no item with that key exists in the list.	var l_list: list(key: it) of int = {5; 4; 3; 2; 1}; print l_list.key_index(2); //prints 3
key_exists (*key-exp*);	The key to be searched for	Returns TRUE if the key exists in the list, or FALSE if it does not.	var l_list: list(key: it) of int = {5; 4; 3; 2; 1}; print l_list.key_exists(2); //prints TRUE

Example 4-6 shows an instance in which the list named *cl* is declared to be a keyed list of four-bit uints, with the key being the list item itself. That is, the key is the value of a four-bit uint. A list of 10 items is built up by generating items and adding them to the keyed list in the for loop.

Example 4-6 Keyed Lists

```
Example of keyed lists
<'
extend sys {
    !cl: list(key: it) of uint(bits: 4);
    run() is also {
        var ch: uint(bits: 4);
        for i from 0 to 10 {
            gen ch;
            cl.add(ch);
        };
```

Example 4-6 Keyed Lists (Continued)

```
        if cl.key_exists(8) then {
            print cl;
            print cl.key_index(8);
        };
    };
};
'>


Results
cl = (10 items, decimal):
            4 5 8 3 14 9 11 4 5 13

cl.key_index(8) = 2
```

We discussed only simple applications of keyed lists. Keyed lists can be used with complex structs, where a field in that struct can be used as a key.

4.5 Creating Struct Subtypes with when

The **when** struct member creates a conditional subtype of the current struct type, if a particular field of the struct has a given value. This is called **when** inheritance, and is one of two techniques *e* provides for implementing inheritance. The other method of creating subtypes is called **like** inheritance.

When inheritance is the recommended technique for modeling in *e*. Like inheritance is more appropriate for procedural testbench programming. Like inheritance is not covered in this book.

4.5.1 When Construct

The purpose of a **when** keyword is to create a **struct** subtype. You can use the **when** construct to create families of objects, in which multiple subtypes are derived from a common base struct type.

A subtype is a struct type in which specific fields of the base struct have particular values, for example:

- If a struct type named "*packet*" has a field named "*kind*" that can have a value of "*eth*" or "*atm*", then two subtypes of "*packet*" are "*eth packet*" and "*atm packet*".

- If the "*packet*" struct has a boolean field named "*good*", two subtypes are "*FALSE'good packet*" and "*TRUE'good packet*".

- Subtypes can also be combinations of fields, such as "*eth TRUE'good packet*" and "*eth FALSE'good packet*".

Struct members defined in a **when** construct can be accessed only in the subtype, not in the base struct. This provides a way to define a subtype that has some struct members in common with the base type and all of its other subtypes, but has other struct members that belong only to the current subtype.

The general syntax for a **when** construct is as follows:

```
when struct-subtype base-struct-type
        {struct-member; …};
```

The applications of the **when** keyword are described in Table 4-8 below.

Table 4-8 When Keyword

base-struct-type	The struct type of the current struct (in which the subtype is being created).
struct-member	Definition of a struct member for the struct subtype. One or more new struct members can be defined for the subtype.
struct-subtype	A subtype declaration in the form *type-qualifier'field-name*.
	The type qualifier is one of the legal values for the field named by field name. If the field name is a boolean field, and its value is TRUE for the subtype, you can omit type qualifier. That is, if "big" is a boolean field, "big" is the same as "TRUE'big".
	The field name is the name of a field in the base struct type. Only boolean or enumerated fields can be used. If the field type is boolean, the type qualifier must be TRUE or FALSE. If the field type is enumerated, the qualifier must be a value of the enumerated type. If the type qualifier can apply to only one field in the struct, you can omit '*field-name*.
	More than one *type-qualifier'field-name* combination can be stated, to create a subtype based on more than one field of the base struct type.

Example 4-7 shows the definition of a **struct** instance where the "op1" field in the struct definition below can have one of the enumerated "*reg_n*" type values (*REG0, REG1, REG2,* or *REG3*). The "kind" field can have a value of "*imm*" or "*reg*", and the "*dest*" field can have a value of "*mm_1*" or "*reg*".

The "*REG0'op1*" subtype specification in the first **when** construct creates a subtype of instances in which the "*op1*" value is "*REG0*". This subtype has all the "*instr*" struct fields plus a "*print_op1()*" method.

The "*reg'kind*" subtype specification in the second **when** construct creates a subtype of instances in which the "*kind*" value is "*reg*". This subtype also has all the "*instr*" struct fields plus a "*print_kind()*" method.

It is necessary to add the "'*kind*" expression in the second **when** construct because the "*dest*" field can also have a value of reg, which means that "*reg*" is ambiguous without the further specification of the field name. This is the explicit definition of a **when** construct.

Example 4-7 Explicit When Construct

```
// Example of explicit when construct
<'
type reg_n : [REG0, REG1, REG2, REG3]; //Define an enumerated type
type instr_type: [imm, reg]; //Define an enumerated type
type dest_type: [mm_1, reg]; //Define an enumerated type

struct instr { // Define a struct
    %op1: reg_n; //This field is always visible
    kind: instr_type; //This field is always visible
    dest: dest_type; //This field is always visible


    when REG0'op1 instr { // The print_op1() method is visible only
                          // when op1 field value is generated to be
                          // REG0. This is an explicit when construct.
        print_op1() is {
            out("instr op1 is REG0");
        };
    };
    when reg'kind instr { // The print_kind() method is visible only
                          // when kind field value is generated to be
                          // reg. This is an explicit when construct.
        print_kind() is {
            out("instr kind is reg");
        };
    };
};
'>
```

Example 4-8 shows an instance of the "*packet*" struct that has a field "*kind*" of either "*transmit*" or "*receive*". The **when** construct creates a "*transmit packet*" subtype. The "*length*" field and the **print**() method apply only to packet instances that have "*kind*" values of "*transmit*".

Example 4-8 Implicit When Construct

```
Example of a when construct
<'
type packet_kind: [transmit, receive]; //Define an enumerated type
struct packet {
    kind: packet_kind; //Define a field of the enumerated type
    when transmit packet { // When the kind field is "transmit"
                           // only then are the additional length
                           // and the print method visible.
                           // Notice that since there is only one
                           // field kind that can have the value
                           // transmit, there is no need to specify
                           // the field kind explicitly
                           // i.e. when transmit'kind packet{.
        length: int; //Field only visible when kind == transmit
        print() is { //Method only visible when kind == transmit
            out("packet length is: ", length);
        };
    };
};
'>
```

4.5.2 Extending When Subtypes

There are two general rules governing the extensions of when subtypes:

- If a struct member is declared in the base struct, it cannot be redeclared in any **when** subtype, but it can be extended.

- With the exception of coverage groups and the events associated with them, any struct member defined in a **when** subtype does not apply or is unknown in other subtypes, including fields, constraints, events, methods, **on**, **expect**, and **assume** constructs.

4.5.3 Extending Methods in When Subtypes

A method defined or extended within a **when** construct is executed in the context of the subtype and can freely access the unique struct members of the subtype with no need for any casting.

When a method is declared in a base type, each extension of the method in a subtype must have the same parameters and return type as the original declaration. In Example 4-9, because *do_op()* is defined with two parameters in the base type, extending *do_op()* in the ADD subtype should also have two parameters only. If the number of parameters is different, a load time error will result.

Example 4-9 Extending Methods Defined in Base Type from When Subtypes

```
Example that shows how methods can be extended in subtypes
Such extensions change the behavior of the methods based
on the values of the fields generated.
If the method is already defined in the base type,
it should have the same number of arguments in the when extension.
<'
struct operation {
    opcode: [ADD, ADD3];
    op1: uint;
    op2: uint;

    do_op(op1: uint, op2: uint): uint is { // Method defined in base
                                           // struct
        result = op1 + op2;
    };
};

extend operation {
    when ADD3'opcode operation { //When opcode == ADD3 additional
                                 //fields and extensions are added.
      op3: uint; //New field in extension
        do_op(op1:uint,op2:uint): uint is also {
        result = result + 2;
      };
    };
};
'>
```

However, if a method is not declared in the base type, each definition of the method in a subtype can have different parameters and return type. The variation shown in Example 4-10 below loads without error.

Example 4-10 Extending Methods Not Defined in Base Type from When Subtypes

```
Example when method that is not defined in the base type but
is defined in the when subtype
<'
struct operation { //Base struct, no method definition
    opcode: [ADD, ADD3];
    op1: uint;
    op2: uint;
};

extend operation {
    when ADD operation { //Define method with 2 arguments
        do_op(op1: uint, op2: uint): uint is {
            return op1 + op2;
        };
    };
    when ADD3 operation { //Define method with 3 arguments
      op3: uint;
      do_op(op1:uint,op2:uint,op3:uint): uint is {
          return op1 + op2 +op3;
      };
    };
};
'>
```

4.6 Units

Originally, all objects in *e* were defined using the **struct** keyword. A problem with this approach was that if a verification environment was built for a certain level of HDL hierarchy, the environment could not be easily ported to another level of HDL hierarchy without significant changes to the *e* code. This led to the introduction of the **unit** keyword in the *e* language.

4.6.1 Unit Overview

Units are the basic structural blocks for creating verification components (verification cores) that can easily be integrated together to test larger and larger portions of an HDL design as it develops. Units, like structs, are compound data types that contain data fields, procedural methods, and other members. However, units are intended as main topology entities. They have some additional features that are not needed for structs.

Unlike structs, a unit instance may be bound to a particular component in the DUT (an HDL path). Furthermore, each unit instance has a unique and constant place (an *e* path) in the runtime data structure of an *e* program. Both the *e* path and the complete HDL path associated with a unit instance are determined during pre-run generation (pre-run generation will be explained in detailed in Chapter 11). Thus a unit instance can be moved to a different level of verification hierarchy by simply changing the HDL path of the unit instance.

4.6.2 Units vs. Structs

The decision of whether to model a DUT component with a **unit** or a **struct** often depends on your verification strategy. Compelling reasons for using a unit instead of a struct include:

- You intend to test the DUT component both standalone and integrated into a larger system. Modeling the DUT component with a unit instead of a struct allows you to use relative path names when referencing HDL objects. When you integrate the component with the rest of the design, you simply change the HDL path associated with the unit instance and all the HDL references it contains are updated to reflect the component's new position in the design hierarchy.

- Your *e* program has methods that access many signals at run time. The correctness of all signal references within units can be determined and checked during pre-run generation, whereas relative HDL references within structs are checked at run time.

On the other hand, using a struct to model abstract collections of data, like packets, allows you more flexibility as to when you generate the data. With structs, you can generate the data during pre-run generation, at runtime, or on the fly, possibly in response to conditions in the DUT. Unit instances, however, can only be generated during pre-run generation. Figure 4-2 shows the differences between structs and units.

Figure 4-2 Comparison between Structs and Units

	Model	Remarks
Units	Environment topology and configuration	• Generated pre-run • Each stub is bound to one DUT interface
Structs	Data item	• Generated during the run, just before being used

4.6.3 Defining Units

Units are the basic structural blocks for creating verification components (verification cores) that can easily be integrated together to test larger designs. Units are a special kind of struct, with two important properties:

- Units or unit instances can be bound to a particular component in the DUT (an HDL path).

- Each unit instance has a unique and constant parent unit (an *e* path). Unit instances create a static tree, determined during pre-run generation, in the run time data structure of an *e* program.

Units are defined with the keyword **unit**. The syntax definition of a **unit** is as follows:

```
unit unit-type [like base-unit-type] {
       [unit-member; …] };
```

The components cited in the above unit definition are explained in further detail in Table 4-9 below.

Table 4-9 Components of a Unit Definition

unit-type	Denotes the type of the new unit.
base-unit-type	Denotes the type of the unit from which the new unit inherits its members.
unit-member; …	Denotes the contents of the unit. Like structs, units can have the following types of members: • data fields for storing data • methods for procedures • events for defining temporal triggers • coverage groups for defining coverage points • **when**, for specifying inheritance subtypes • declarative constraints for describing relationships between data fields • **on**, for specifying actions to perform upon event occurrences • **expect**, for specifying temporal behavior rules Unlike structs, units can also have **verilog** members. The definition of a unit can be empty, containing no members.

Example 4-11 illustrates a simple **unit** definition.

Example 4-11 Basic Unit Definition

```
Example of unit definition
<'
unit router_channel { //Definition of unit router_channel
    //Examples of struct members that can be defined in a unit
    //At this point, the syntax of these struct members is not relevant
    event external_clock;
    event packet_start is rise('valid_out')@sim;
    event data_passed;

    verilog variable 'valid_out' using wire;

    data_checker() @external_clock is {
        while 'valid_out' == 1 {
        wait cycle;
        check that 'data_out' == 'data_in';
        };
    emit data_passed;
    };

    on packet_start {
        start data_checker();
    };
}; //End of unit definition

'>
```

4.6.4 HDL Paths

Relative HDL paths are essential in creating a verification module that can be used to test a DUT component either standalone or integrated into different or larger systems. Binding an *e* unit instance to a particular component in the DUT hierarchy allows you to reference signals within that DUT component using relative HDL path names. Regardless of where the DUT component is instantiated in the final integration, the HDL path names are still valid.

To illustrate this concept, let's look at the *fast_router* shown in Figure 4-3. This figure shows the *e* hierarchy with **sys** at the topmost level.

Figure 4-3 *e* Hierarchy of the fast_router

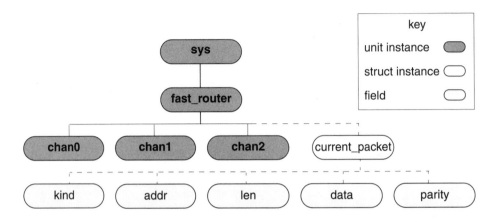

Each unit instance in the unit instance tree of the fast_router matches a module instance in the Verilog DUT, as shown in Figure 4-4. The one-to-one correspondence in this particular design between *e* unit instances and DUT module instances is not required for all designs. In more complex designs, there may be several levels of DUT hierarchy corresponding to a single level of hierarchy in the tree of *e* unit instances.

Figure 4-4 DUT Router Hierarchy

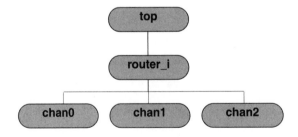

To associate a unit or unit instance with a DUT component, you use the **hdl_path**() method within a **keep** constraint. Example 4-12 extends **sys** by creating an instance of the *XYZ_router* unit and binds the unit instance to the *"router_i"* instance in the DUT.

Example 4-12 HDL Path for the fast_router

```
Example of a fast_router with hdl_path() binding
to an HDL hierarchy. All simulator signal accesses
are now done relative to the hdl_path()
<'
unit fast_router { //Define a fast_router unit. This is only a
                   //simple example. Actual fast_router code will
                   //contain lots of verification code.
    debug_mode: bool;
};

extend sys {
    unit_core: fast_router is instance; //Instantiate the fast_router
    keep unit_core.hdl_path() =="top.router_i"; //Associate this
                                        // instance with the HDL design
    keep unit_core.debug_mode == TRUE; //Constraint to set debug mode
};
'>
```

Similarly, Example 4-13 creates three instances of *router_channel* in *fast_router* and constrains the HDL path of the instances to be *"chan0"*, *"chan1"*, *"chan2"*. These are the names of the channel instances in the DUT relative to the *"router_i"* instance.

Example 4-13 HDL Path for the Channels

```
Example showing how the channels are instantiated
inside the fast_router and the assignment of hdl_path()
for those channels.
<'
unit router_channel { //Code in this unit not shown
};

unit fast_router { //Definition of fast router
    channels: list of router_channel is instance;//Instantiate
                                            //channels
    keep channels.size() == 3; //There are 3 channels
    keep for each in channels { //Set the hdl_path() for each channel
      .hdl_path() == append("chan", index); };
};
'>
```

The full HDL path of each unit instance is determined during pre-run generation, by appending the HDL path of the child unit instance to the full path of its parent, starting with the **sys** hierarchy. **sys** has the empty full path "". Thus the full path for the *fast_router* instance is "*top.router_i*" and that for the first channel instance is "*top.router_i.chan0*".

4.6.5 Predefined Methods for Units

Besides **hdl_path**() there is another predefined method for units called **get_enclosing_unit**() that is commonly used.

4.6.5.1 get_enclosing_unit()

This predefined method of a **unit** returns a reference to the nearest higher-level unit instance of the specified type, allowing you to access fields of the parent unit in a typed manner. The syntax for this predefined method is as follows:

```
[exp.]get_enclosing_unit(unit-type: exp): unit instance;
```

The arguments to this method are shown in Table 4-10 below.

Table 4-10 Arguments for get_enclosing_unit() Method

exp	An expression that returns a unit or a struct. If no expression is specified, the current struct or unit is assumed.
unit-type	The name of a unit type or unit subtype.

Example 4-14 shows the usage of this method.

Example 4-14 Usage of get_enclosing_unit()

```
unpack(p.get_enclosing_unit(fast_router).pack_config,
       'data', current_packet);
```

4.6.5.2 Other Predefined Methods

Many other predefined methods are available with unit definitions. Please refer to the *e* Language Reference Manual for details.

4.7 Summary

We discussed the concepts of structs and units in this chapter. These concepts lay the foundation for the material discussed in the following chapters.

- Structs are defined using the keyword **struct**. Struct definitions contain struct members.

- Structs can be extended using the keyword **extend**. Struct extensions add struct members to a previously defined **struct**. These extensions can be specified in the same file or a different file.

- List fields are used very frequently in *e*. Lists of any data type, enumerated type, or user-defined struct can be defined in *e*.

- Many pseudo-methods are available to operate on lists. When a list field is defined, the pseudo-methods can be used with that list field. Any arguments required by the pseudo-method go in parentheses after the pseudo-method name. Pseudo-methods can be called in actions or constraints.

- Keyed lists are used to enable faster searching of lists by designating a particular field or value which is to be searched for. Although all of the operations that can be done using a keyed list can also be done using a regular list, using a keyed list provides an advantage in the greater speed of searching a keyed list.

- The **when** struct member creates a conditional subtype of the current struct type, if a particular field of the struct has a given value. This is called "when" inheritance, and is one of two techniques *e* provides for implementing inheritance. The other method is called **like** inheritance. When inheritance is the recommended technique for modeling in *e*. Like inheritance is more appropriate for procedural testbench programming. Like inheritance is not covered in this book.

- When subtypes can be extended.

- A method defined or extended within a **when** construct is executed in the context of the subtype and can freely access the unique struct members of the subtype with no need for any casting.

- The keyword **unit** is used to define static components in the verification environment. On the other hand, using a **struct** keyword to model abstract collections of data, like packets, allows you more flexibility as to when you generate the data.

- HDL paths can bind an *e* unit instance to a particular component in the DUT hierarchy. This allows you to reference signals within that DUT component using relative HDL path names. HDL paths can be assigned using the **hdl_path**() method.

4.8 Exercises

1. Define a **struct** named "*instruction*" that contains the following fields:

    a. *opcode*: 8 bits
    b. *op1*: 8 bits
    c. *op2*: 8 bits

2. Extend the definition of **struct** *"instruction"* by adding the following fields:

　a. *kind_op2*: A field of enumerated type with two values, *imm* and *reg*. (Hint: You may need to define an enumerated type before you add the field.)

　b. *memory_instr*: boolean

3. Extend the definition of the predefined **struct sys** by adding the following fields:

　a. *l_instructions*: A list of struct *instruction*

　b. *data*: A list of 8-bit unsigned integers

　c. *memory_test*: boolean

4. Extend the **run()** method within sys to modify the *l_instructions* list as follows. (Hint: You may need to add a variable of type instruction before modifying the list.)

　a. Add an instruction to the end of the list.

　b. Delete an instruction from the head of the list.

　c. Insert an instruction at index == 5.

5. Extend the definition of **struct** *"instruction"* using the when subtype to conditionally add the following fields:

　a. When the field *kind_op2* is generated as *imm*, define a new field called *imm_value* that is an 8-bit unsigned value.

　b. When the field *kind_op2* is generated as *reg*, define a new field called *reg_value* that is an 8-bit unsigned value.

6. For each of the following, choose whether a **struct** or a **unit** should be used to represent it in *e*.

　a. A data object used to represent a single test vector

　b. An input driver that drives each data object onto the device

　c. A protocol monitor that monitors the protocol on the bus

7. Define a **unit** named *"port"* that contains the following fields:

　a. *port_id*: 8 bits

　b. *data*: 8 bits

　c. *clock*: 8 bits

8. Create three instances (*port0*, *port1*, and *port2*) of the **unit** port in **struct sys**.

9. In **struct sys** bind the instances of unit *"port"* to the corresponding HDL hierarchy instances using **hdl_path()** as follows:

　a. *port0* to *"~/top/router_i/chan0"*

　b. *port1* to *"~/top/router_i/chan1"*

　c. *port2* to *"~/top/router_i/chan2"*

10. Load all code from exercises 2 to 9 into Specman Elite and verify that it loads without error.

Constraining Generation

The purpose of defining structs and units is to build a verification hierarchy. However, each instance of a struct or unit and their fields must be generated in a manner that is compliant with the design specification and that follows the test plan. Constraints are used to control generation of structs and units. This chapter discusses how constraints are applied to structs and units to generate a verification environment that is meaningful.

Chapter Objectives

- Describe basic concepts of constraints and generation.
- Explain simple constraints.
- Explain implication constraints.
- Describe soft constraints.
- Understand weighted constraints.
- Describe how to extend **struct** subtypes.
- Explain order of generation.
- Explain constraint resolution mechanisms.
- Understand do-not-generate fields.

5.1 Basic Concepts of Generation

Generation is the process that automates the selection of values for fields and variables (data items). During the generation phase, the entire tree of instances under the struct **sys** is generated. Data values are assigned to all fields. In Figure 5-1, values will be generated for fields in instances *driver1*, *checker1*, *receiver1*, *data1*, *protocol1*, *error1*, and *collect1* in a depth-first manner.

Figure 5-1 Generation Hierarchy

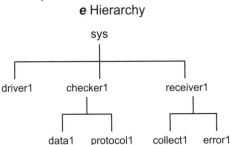

Constraints are directives that influence the behavior of the generator. They are struct members that influence the generation of values for data items within the struct and its sub-tree. There are two basic types of constraints:

1. Value constraints restrict the range of possible values that the generator produces for data items, and they constrain the relationship between multiple items.

2. Order constraints influence the sequence in which data items are generated. Generation order is important because it affects the distribution of values and the success of generation.

Both value and order constraints can be hard or soft:

- Hard constraints (either value or order) must be met or an error is issued.
- Soft value constraints suggest default values but can be overridden by hard value constraints.
- Soft order constraints suggest modifications to the default generation order, but they can be overridden by dependencies between data items or by hard order constraints.

You can define constraints in many ways:

- By defining a range of legal values in the field or variable declaration.
- By specifying values or attributes of items in a list.
- By using one of the constraint variations within a struct definition.
- By using an on-the-fly constraint within a method.

By default, generation takes place before the simulator is invoked. However, you can generate values for particular struct instances, fields, or variables during simulation with on-the-fly generation.

5.1.1 Need for Constraints

If stimulus is applied using completely random values, the results will not be meaningful. Constraints are needed because a large part of the design is exercised only by stimuli that have specific legal relationships between the fields. These legal relationships are set by the design specification and the test plan.

Figure 5-2 shows that when a design is presented with an illegally constructed data object, it will typically discard it without involving most of the logic in the design. Our verification goal is to exercise the entire design, so we need to make sure that most data objects meet the requirements to "cross the wall" to the majority of the logic. These requirements can be expressed as necessary relationships between object field values.

Figure 5-2 Applying Constrained Objects on a Design

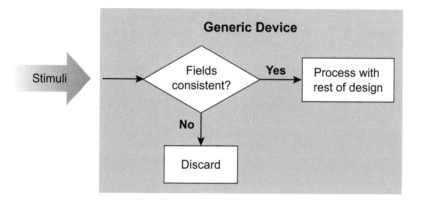

5.1.2 Examples of Constraints

Examples of constraints in a verification system are as follows:

- In a data packet, the value in the type field might affect the number of bytes allowed in the payload.
- Configuring a DUT requires constraining the bus transaction address.
- For particular opcodes, some CPU registers are invalid targets.
- A specific bus protocol only supports burst sizes of 1, 4, 8, or 16.

5.1.3 Generic Constraint Solver

A generic constraint solver collects all the constraints in the *e* code at any level of hierarchy, resolves these constraints, and generates the data for all structs within the struct **sys** hierarchy. Figure 5-3 shows how a constraint solver is provided as a part of Specman Elite. A data object is a struct that represents one test vector. A port object represents the interfaces that drive the data object on to the DUT. The system config object holds system level information. Specman Elite takes the constraints from all these objects and solves them to assign valid values to the fields in the various instances.

Figure 5-3 Generic Constraint Solver

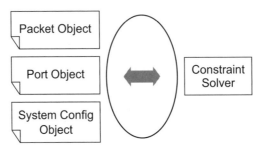

5.1.4 Generation Mechanism

For each field in the verification environment, the generator tries to satisfy all constraints that are specified. If a solution to the constraints is found, a value is randomly picked from the range of legal solutions and assigned to the generated field. If such a solution is not found, a contradiction error is displayed. This process is done for each field in the environment. Figure 5-4 shows the decision chart for generation of values for fields.

Figure 5-4 Generation Decision Chart

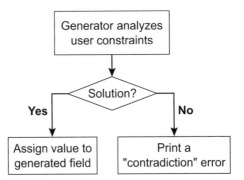

5.1.5 Directed-Random Simulation

The *e* language allows a verification engineer to have full control of the values generated for the fields during simulation. If the values are constrained to very narrowly defined values, it is a directed test. If the values are completely unconstrained, it is a random test. If the values are constrained to a range, then it is a directed-random test. As more constraints are added, the simulation becomes more directed. Thus, an engineer can exert full control over the simulation by constraining stimulus generation and environment variables appropriately. Table 5-1 shows the different simulation scenarios for a 32-bit *address* field.

Table 5-1 Directed-Random Testing

Test Type	Example	Description
Random	address: uint(bits:32);	No constraint on address field. Random testing.
Directed-random	keep address in [1000..2000];	Address is constrained to a range of values. Value within the range is chosen randomly. Directed-random testing.
Directed	keep address == 1200;	Address is constrained to a specific value. Directed testing.

5.2 Basic Constraints

A constraint is a boolean expression that specifies the relationships between various fields. The most basic form of a constraint is specified using the keyword **keep**. The syntax for definition of constraints is as follows:

```
keep constraint-bool-exp;
```

A constraint is a struct member. Example 5-1 shows how a simple constraint can be specified to define relationships between fields.

Example 5-1 Simple Constraint Definition

```
Simple example of constraint definition
<'
struct pkt {
    kind: [tx, rx];
    len: uint;
    keep kind != tx or len == 16;
};
'>
```

Example 5-2 shows other variations of constraints. This example assumes that the fields have already been defined in a previous struct definition.

Example 5-2 More Constraint Definitions

```
Example provides more definitions of constraints.
The constraints assume that the fields in the struct
my_struct have been defined previously in the original
definition of the struct. The constraints simply control
the generation of the fields.
<'
extend my_struct{
   keep x > 4 and x != 6; //Can use and, or, not constructs
   keep x == y + 25; // Can use addition and subtraction
   keep z == TRUE; // z is constrained to be always true
   keep y == method(); // Can include method calls
   keep x in [5..10,20..30]; // Can constrain variable to subranges
   keep x not in [1..3, 5..8]; // Not in a set of subranges
   keep for each (p) in packets{ // Set constraint for each item in
                                 // list. p is just a placeholder.
                                 // packets is a list of structs.
       p.length < 10; // The length field for each packet < 10
   };
   keep packets.size() == 50; //Can use list pseudo-methods in calls
};
'>
```

If two constraints contradict each other, then no value can be assigned to the field and a contradiction error is displayed.

5.3 Implication Constraints

Implication constraints set up constraints on fields or list members based on another boolean expression being true. Implication constraints are specified using the **=>** operator (Note that X => Y is equivalent to *not X or (X and Y)*). Example 5-3 shows how the implication operator can be used in the context of a list. Each member of the list can be constrained separately based on the implicit *index* operator.

Example 5-3 Implication Constraints

```
Example to show implication constraints on a single field
as well as for constraining a list.
<'
type packet_size_t : [SHORT, MEDIUM, LONG]; //Defined enumerated type
```

Example 5-3 Implication Constraints (Continued)

```
struct packet { // Example shows implication constraint on field.
    size: packet_size_t; //Define a field of type packet_size_t.
                         //The generator picks a value for size.
                         //The size field is defined to control the
                         //generation of the len field.

    address: uint(bits:8); //Define an address 8 bits wide
    len: uint(bits:8); //Define a len field 8 bits wide
        keep size == SHORT => len < 10; //If size value picked is
                                        //SHORT, then a constraint
                                        //len < 10 applies.
        keep size == MEDIUM => len in [11..19];
                                        //If size value picked is
                                        //MEDIUM, then a constraint
                                        //len in [11..19] applies.
        keep size == LONG => len < 20; //If size value picked is
                                        //SHORT, then a constraint
                                        //len < 20 applies.

}; //End of struct packet

extend sys {
    packets: list of packet;
        keep for each (p) in packets {//Implication constraint on list
                                      //items, index is an implicit
                                      //iterator value.
            index == 0 => p.addr == 0; //For first packet, addr == 0.
            index == 1 => p.addr == 1; //For second packet, addr == 1.
            index > 2 => p.addr in [2..15]; //For other packets,
                                             // addr is in [2..15].
        };
}; //End of sys
'>
```

5.4 Soft Constraints

Often constraints are specified as a default preference. These constraints need to be overridden later. These are called *soft constraints* and are specified using the keywords **keep soft**. Soft value constraints on a data item are considered only at the time the data item is generated, after the hard value constraints on the data item are applied. Soft constraints are evaluated in reverse order of definition. If a soft constraint conflicts with the hard constraints that have already been applied, it is skipped. If there is a contradiction with other soft constraints, the last loaded soft constraint prevails.

In Example 5-4 below, the constraints will be evaluated in the following order:

1. The hard constraint is applied, so the range is $[1..10]$.

2. The last soft constraint in the code order, $x < 6$, is considered. It does not conflict with the current range, so it is applied. The range is now $[1..5]$.

3. The next to last soft constraint, $x == 8$, conflicts with the current range, so it is skipped. The range is still $[1..5]$.

4. The first soft constraint in the code order, $x > 3$, does not conflict with the current range, so it is applied. The final range of legal values is $[4, 5]$.

Example 5-4 Evaluation Order of Soft Constraints

```
Example shows the order in which the constraints are evaluated
<'
struct cons {
   x: uint;
   keep x in [1..10]; //First eval, range of x is [1..10]
   keep soft x > 3;    //Fourth eval, no conflict range of x is [4, 5]
   keep soft x==8;     //Third eval, conflict with [1..5], skipped
   keep soft x < 6;    //Second eval, no conflict, range of x is [1..5]
};
'>
```

Example 5-5 shows more examples of soft constraints.

Example 5-5 Soft Constraints

```
Example shows different combinations of soft constraints
and hard constraints and the corresponding results.
<'
struct example {
   len1: uint;
     keep soft len1 == 64; //No contradiction with hard constraint
     keep len1 <= 100; //Result is len1 equal to 64

   len2: uint;
     keep soft len2 == 64; //Contradiction with hard constraint
     keep len2 >= 100; //Result is len2 value >= 100

   len3: uint;
     keep soft len3 > 64;
     keep soft len3 < 64; //Contradiction with previous soft
                          //constraint. Last loaded constraint
                          //prevails i.e. len3 > 64
};
'>
```

5.4.1 Resetting Soft Constraints

A soft constraint usually indicates the default preferences for a certain field. In order to override the default and change the value with a hard constraint, the verification engineer must also reset the soft constraints for that field using the **reset_soft()** constraint. It is advisable to do this explicitly to make sure that there are no residual soft constraints acting on that field. Example 5-6 shows how to reset soft constraints.

Example 5-6 Resetting Soft Constraints

```
Example shows a soft constraint is reset before another hard
constraint is applied in the extension of the struct.
<'
struct packet { //Original struct definition with soft constraint.
   len: uint;
   keep soft len in [64..1500]; //Default range
};

extend packet { //Struct extension, typically in separate file
              //but shown here in the same file for convenience.
    keep len.reset_soft(); //Reset all previously loaded soft
                           //constraints on the len field.
                           //reset_soft() is on a per field basis.
    keep len > 2000;       //Apply a hard constraint.

};
'>
```

5.5 Weighted Constraints

Weighted constraints are soft constraints used to specify a non-uniform distribution of values. A weighted constraint specifies the relative probability that a particular value or set of values is chosen from the current range of legal values. Weighted constraints are specified using the keywords **keep soft select**. The current range is the range of values as reduced by hard constraints and by soft constraints that have already been applied.

A weighted value will be assigned with the probability of *weight/(sum of all weights)*. Weights are treated as integers. Example 5-7 shows the specification of weighted constraints.

Example 5-7 Weighted Constraints

```
Example shows the distribution for opcodes in
an instruction.
<'
type cpu_opcode: [
    ADD, ADDI, SUB, SUBI,
    AND, ANDI, XOR, XORI,
    JMP, JMPC, CALL, RET,
    NOP
] (bits: 4); //Define opcode enumerated type

struct instr {
    opcode: cpu_opcode; //Field opcode
    keep soft opcode == select {
        30: ADD; // 30/60 probability that ADD is selected
        20: ADDI; // 20/60 probability that ADDI is selected
        10: [SUB..NOP]; // 10/60 probability SUB..NOP are selected
        //Note that the weights do not have to add up to 100.
        //They are simply relative weights.
    };
}; //End of struct instr
'>
```

5.6 Order of Generation

Specman Elite generates all instances instantiated in the **sys** hierarchy. It generates the values of the fields in different instances of the struct in a *depth-first* order. Figure 5-5 shows sample *e* code and its corresponding order of generation. The list *packets* is generated first. The generator goes depth-first into each packet and generates the values for each field in the order they are defined. The numbers shown in Figure 5-5 explain the order in which the fields and structs are generated.

Figure 5-5 Order of Generation

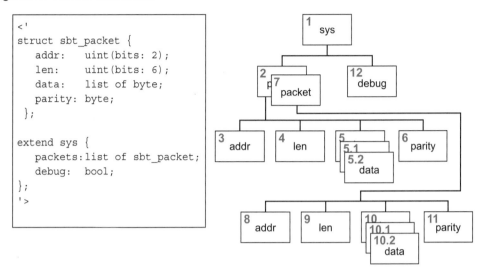

The default order of generation within a **struct** is the order in which the fields are defined. However, there are special cases in which the generation order can be changed either explicitly or implicitly. The following sections discuss such cases.

5.6.1 Implicit Generation Order

Certain constraints imply a specific generation order. Table 5-2 shows certain constraints that imply specific generation order.

Table 5-2 Constraints that Imply Generation Order

Description	Example	Generation Order
Constraints involving a method call	keep parity == calc_parity(data);	*data* is generated before *parity*
List slicing	keep z == my_list[x..y];	*x* and *y* are generated before *z*
Multiply, divide, and modulo	keep z == x*y;	*x* and *y* are generated before *z*

5.6.2 Explicit User-Defined Generation Order

The default generation order is the order in which fields are defined in a struct. One can override this default order by adding explicit order rules using the **keep gen before** constraint. Example 5-8 below shows how to override the default generation order.

Example 5-8 Explicit User-Defined Generation Order

```
Example shows generation order is explicitly defined
using the keep gen before syntax. This syntax is especially
useful with implication constraints.
<'
struct packet { //Original struct definition
   addr: uint;
   len: uint(bits:4); //Default generation order is addr before len
};

extend packet { //Extension that could be in a different file
    keep len == 5 => addr < 50; //To work correctly this implication
                                //must have len generated before addr
    keep gen (len) before (addr); //Ensures the len is generated
                                //before addr.
};
'>
```

5.7 Constraint Resolution

While generating fields, the generator must satisfy all relevant constraints. These constraints may be defined in several places including:

- The original **struct** definition
- Extensions to the **struct** definition
- Other **struct** definitions using the hierarchical path notation

We have seen many examples of constraints in the original **struct** definition and the extensions to the **struct** definition. Example 5-9 shows a constraint defined using the hierarchical path notation.

Example 5-9 Resolution of Hierarchical Constraints

```
Example shows the resolution of a constraint
specified using a hierarchical path notation.
<'
struct cell { //Original struct definition
    len: uint(bits:8); //Field len, default range [0..255]
    keep len in [0..48]; //All cells constrain len to [0..48]
};

struct port {
    cell_inst: cell; //Instantiate the cell
    keep cell_inst.len > 5; //Hierarchical constraint for cell_inst
                            //len is now constrained to [6..48]
};
'>
```

5.8 Do-Not-Generate Fields

All fields within the **sys** hierarchy are generated by Specman Elite. However, if certain fields are going to be generated or collected procedurally later in the simulation, they should not be generated. One can inhibit the generation of such fields by prefixing these fields with a **!** symbol. Example 5-10 shows an example of do-not-generate fields.

Example 5-10 Do-Not-Generate Fields

```
Example shows the specification of do-not-generate fields.
<'
struct cell {
    len: uint(bits:8); //Field len will be generated if
                        //cell instance is generated.
};

struct port {
    !cell_inst: cell; //Instantiate the cell but do not generate
    !count: uint; //Define field count, but do not generate value
};
'>
```

5.9 Summary

- Generation is the process that produces values for fields and variables (data items). During the generation phase, the entire tree of instances under the struct **sys** is generated.
- Constraints are needed because a large part of the design is exercised only by stimuli that have specific legal relationships between the fields. These legal relationships are set by the design specification and the test plan. Constraints are directives that influence the behavior of the generator. They are declared within a struct and influence the generation of values for data items within the struct and its sub-tree.
- There are two basic types of constraints: *value* constraints and *order* constraints.
- A constraint is a boolean expression that specifies the relationships between various fields. The most basic form of a constraint is specified using the keyword **keep**.
- Implication constraints set up constraints on fields or list members based on another boolean expression being true. Implication constraints are specified using the => operator.
- Often constraints are specified as a default preference. These constraints need to be overridden later. These are called *soft constraints* and are specified using the keywords **keep soft**.
- All soft constraints for a field can be reset with the **reset_soft()** constraint.
- A weighted constraint specifies the relative probability that a particular value or set of values is chosen from the current range of legal values. Weighted constraints are specified with the keywords **keep soft select**.
- The default generation order is the order in which fields are defined in a **struct**. One can override this default order by adding explicit order rules using the **keep gen before** constraint.
- One can inhibit generation of fields by prefixing these fields with a ! symbol.

5.10 Exercises

1. Explain the differences between value constraints and order constraints.

2. Explain the need for constraints.

3. Explain the differences between random testing and directed-random testing.

4. Define a **struct** named "*instruction*" that contains the fields: *opcode*: 8 bits, *op1*: 8 bits and *op2*: 8 bits. Define the following constraints.

  a. *opcode* should be in the ranges 5..10 and 20..25.
  b. *op1* should be in the range 10..100.

c. *op2* should not be in the ranges 3..8 and 100..200.

d. *op1* should not be equal to *op2* times 2.

e. If *opcode* is 5, then *op1* should be in the range 20..45.

f. If *opcode* is 8, then *op2* should be in the range 150..160.

5. Extend the definition of **struct** *instruction* by adding a field *kind_op2* that is a field of enumerated type with two values, *imm* and *reg*. (Hint: You may need to define an enumerated type before you add the field.) Define the following constraints:

a. If *kind_op2* is *imm*, the *op2* field should be restricted to the range 120..180.

b. Add an explicit user-defined order constraint that generates *kind_op2* before *op2*. (Hint: Use **keep gen before**.)

6. Extend the definition of the predefined **struct sys** by adding the field *l_instructions* to be a list of struct *instruction*. For each instruction in the l_instructions list, define a weighted constraint as follows:

a. Weight =4, opcode = 5.

b. Weight =6, opcode =8.

7. Define a **struct** *packet* that contains the field *len* of the type **byte** and a field *addr* of the type **byte**. For the field len, define a weighted constraint as follows:

a. Weight = 7, *len* is in the range 0..15.

b. Weight = 3, *len* is in the range 16..63.

c. Weight = 10, *len* == 64.

d. Soft constraint on field *addr* to be in the range 0..7.

8. Extend the definition of **struct** *packet* and do the following:

a. Reset the soft constraint on *len* field.

b. Add a constraint for *len* > 63.

c. Reset the soft constraint on *addr* field.

d. Add a constraint for *addr* in range 8..15.

9. Extend the definition of the predefined struct **sys** by adding the field *l_packets* to be a list of *packet*. Define the following constraints in the struct sys for list *l_packets*.

a. First packet should have *addr* ==8.

b. Second packet should have *addr* == 9.

c. Remaining packets should have addr in the range 10..15.

10. Load all code from exercises 4 to 9 into Specman Elite and verify that it loads without error.

CHAPTER 6

Procedural Flow Control

The purpose of defining structs and units is to build a verification hierarchy. To interact with the device under test, procedures are required to drive and sample the signals at the appropriate time. Procedures are also required to create interfaces, compute values, and act upon the fields of a struct or unit. This chapter discusses procedural flow control in *e*. Note that, unlike a verification environment implemented in an HDL, an *e* verification environment is largely declarative. Procedural code is found primarily in bus functional models and scoreboards.

Chapter Objectives

- Describe methods and invocation of methods.

- Explain local variables.

- Explain conditional actions such as if-else and case.

- Describe loops such as while, for, repeat, for each loop.

- Understand predefined actions and methods such as outf, out, pack, and unpack.

6.1 Defining Methods

As discussed in "Syntax Hierarchy" on page 49, *e* contains a very strict syntax hierarchy. Procedural actions can only be defined inside methods. Methods are struct members. *e* methods are similar to C functions, Verilog tasks, and VHDL processes. An *e* method is an operational procedure containing actions that define its behavior. A method can have parameters, local variables, and a return value. You can define a method only within a struct and you must create an instance of the struct before you can execute the method. The syntax to define a method is as follows:

```
method-name ([parameter-list]) [: return-type] is {action;…};
```

The components of a method definition are as shown in Table 6-1 below:

Table 6-1 Parameters of a Method

method-name	A legal *e* name.
parameter-list	A list composed of zero or more parameter declarations of the form *param-name*: *param-type* separated by commas. The parentheses around the parameter list are required even if the parameter list is empty.
	param-name — A legal *e* name.
	param-type — Specifies the parameter type.

Example 6-1 shows a method definition.

Example 6-1 Method Definition

```
Example that shows a simple method definition
<'
struct packet { //Struct definition
 addr: uint(bits:2); //Field of a struct member
 zero_address (): bool is { //Define method name zero_address
                           //No arguments, return type boolean
 if(addr == 0) then {
   out("Packet has address 0");
   result = TRUE; //An implicit variable result is automatically
                  //created and is of the same type as the return
                  //which in this case is bool.
      } else {
      result = FALSE;
 }; //end of if-else
 //The value of result when it reaches the end of the method call
 //is the return value of the method.
}; //end of method definition
'>
```

Methods have the following characteristics:

• Methods can contain actions including variable declarations.
• Methods can have zero or up to 14 input parameters.
• Methods may or may not return a value.

- Methods may or may not consume time. (The methods discussed in this chapter do not consume time.)

6.1.1 Values Accessed in a Method

Methods can read or write the following values:

- Locally declared variables within the method.
- Fields within the local struct.
- Arguments and return value (implicit **result** variable) of the method.
- Fields in other structs using path notation.

Example 6-2 shows various fields that can be accessed in a method.

Example 6-2 Values Accessed in a Method

```
Example shows the different values that can be accessed
inside a method
- Fields within the local struct - len
- Arguments - min_len
- Local variable - count
- Implicit result variable - result
- Hierarchical path notation - d_struct_i.addr
<'
struct c_struct { //Struct definition
  len: uint; //Field len
  legal: bool; //Field legal
  d_struct_i: d_struct; //Field struct instantiation
  legal_length(min_len: uint):bool is {
    var count: int; //Local variable has a default value of 0.
    if (len >= min_len) then { //Directly access len, a field in
                               //c_struct. min_len is an argument.
      result = TRUE;
      count += 1; //Increment local variable count by 1.
    } else {
      result = FALSE;
    };
```

Example 6-2 Values Accessed in a Method (Continued)

```
    if (count == 1)then { //Check the value of the local variable.
     d_struct_i.addr = 0; //Set the value of the addr field in
                          //d_struct_i instance to be 0. Access
                          //field in another struct.
     out("Incremented counter");
    };
  }; //end of method legal_length
}; //end of c_struct definition

struct d_struct {//Define d_struct
  addr: uint;
}; //end of d_struct definition
'>
```

6.1.2 Local Variables

Local variables are declared within a method using the **var** action. Local variables can only be accessed within that method. Local variables can have the same types as fields. Default initial values of local variables are shown in Table 6-2.

Table 6-2 Local Variable Default Values

Variable Type	Default Value	Example
int/uint	0	`var count: uint;`
list	empty	`var b_list: list of byte;`
bool	FALSE	`var legal: bool;`
struct	NULL	`var tmp_packet: packet = new;` (The keyword **new** or **gen** is needed for **var** declaration so that the fields of the struct can be accessed.)

6.1.3 Invoking Methods

Methods are not executed unless they are invoked. One invokes a method by simply calling the method in a predefined or a user-defined method. Example 6-3 shows how the method *legal_length* defined in Example 6-2 is invoked from a predefined **post_generate**() method of the struct *c_struct*.

Example 6-3 Method Invocation

```
Example shows how the method legal_length is invoked
in the predefined post_generate() method. The post_generate()
method is explained later in the book.
<'
extend c_struct { //struct definition
  post_generate() is also {
  legal = legal_length(64); //legal gets the return value of method
                            //64 is the min_len arg passed to method
  };
}; //end of c_struct extend
'>
```

6.1.4 Extending Methods

Similar to structs and units, methods can also be extended. Methods are commonly extended in three ways:

- **is first** extension adds code before the existing method definition.
- **is also** extension adds code after the existing method definition.
- **is only** extension replaces the existing method definition.

Example 6-4 shows an instance in which all three extensions of the method are used. Existing method definition means the code loaded thus far.

Example 6-4 Extending Methods

```
Example shows how to extend methods using the
three extensions.
<'
struct meth {
    m() is { //Original method definition
        out("This is...");
    };
};

extend meth {
    m() is also { //Add code to end of existing definition
        out("This is also..."); //Prints after "This is...".
    };
};

extend meth {
    m() is first { //Add code to beginning of existing definition
        out("This is first..."); //Prints before "This is...".
    };
};
'>
Comment block:
At this point the method looks like the following
    m() is {
        out("This is first...");
        out("This is...");
        out("This is also...");
    };

<'
extend meth {
    m() is only { //Replace existing definition
        out("This is only...");
    };
};
'>

Comment block:
At this point the method looks like the following
    m() is {
        out("This is only...");
    };
```

6.2 Conditional Actions

Conditional actions are used to specify code segments that will be executed only if a certain condition is met.

6.2.1 If-then-else

The syntax of **if-then-else** is shown below:

```
if bool-exp [then] {action; ...}
        [else if bool-exp [then] {action; ...}] [else {action; ...}];
```

If the first *bool-exp* is TRUE, the **then** *action* block is executed. If the first *bool-exp* is FALSE, the **else if** clauses are executed sequentially: if an **else if** *bool-exp* is found that is TRUE, its **then** *action* block is executed; otherwise the final **else** *action* block is executed. Example 6-5 shows various if-then-else actions. Note that the **then** keyword is optional but recommended.

Example 6-5 If-then-else Actions

```
Example showing various if-then-else actions
<'
struct test1 {
    a: int;
    b: int;

    meth1() is { //Define a method called meth
     if a > b then { //then keyword is optional
        print a, b;
     } else {
        print b, a;
     };
    }; //end of meth1 definition
}; //end of struct test1 definition
```

Example 6-5 If-then-else Actions (Continued)

```
struct test2 {
    a_ok: bool;
    b_ok: bool;
    x: int;
    y: int;
    z: int;

    meth2() is {
     //Complex if-else-if clause
     if a_ok { //Note that then keyword is optional
        print x;
     } else if b_ok {
        print y;
     } else {
        print z;
     };
    }; //end of meth2 definition
}; //end of struct test2 definition
'>
```

6.2.2 Case Action

The syntax to define a **case** action is as follows:

```
case case-exp { label-exp : action-block;...};
```

The **case** action executes an action block based on whether a given comparison is true. The **case** action evaluates the *case-exp* and executes the first *action-block* for which *label-exp* matches the *case-exp*. If no *label-exp* equals the *case-exp*, the *case* action executes the *default-action* block, if specified. After an *action-block* is executed, Specman Elite proceeds to the line that immediately follows the entire **case** statement. Example 6-6 shows various case actions.

Example 6-6 Case Action

```
Example shows case actions
<'
struct packet {
    length: int;
};
```

Example 6-6 Case Action (Continued)

```
struct temp {
    packet1: packet; //Instantiate the packet struct

    meth() is {
    case packet1.length { //Case action
     64:          {out("minimal packet")}; //label-exp and action block
    [65..256]:   {out("short packet")};
    [256..512]:  {out("long packet")};
    default:     {out("illegal packet length")};
    };
 }; //end of method meth()
}; //end of struct temp

'>
```

6.3 Iterative Actions

There are four types of iterative actions in *e*:

- For loop
- For each loop
- While loop
- Repeat loop

This section discusses these looping actions.

6.3.1 For Loop

The syntax of an *e* style **for** loop is as follows:

```
for var-name from from-exp [down] to to-exp [step step-exp]
                                        [do] {action;...};
```

A **for** loop executes for the number of times specified by **from to** keywords. The **for** loop creates a temporary variable *var-name* of type **int**, and repeatedly executes the *action* block while incrementing (or decrementing if **down** is specified) its value from *from-exp* to *to-exp* in interval values specified by *step-exp* (defaults to 1). In other words, the loop is executed until the value of *var-name* is greater than the value of *to-exp* (or less than the value of *to-exp* if **down** is used).

Example 6-7 shows various types of the for loop.

Example 6-7 For Loop

```
Example shows different variations of a for loop
<'
struct temp {
    a: int;
    meth() is {
       for i from 2 to 2 * a do { //Simple for loop
                                  //No need to declare i in an e
                                  //style for loop. Variable i is
                                  //automatically declared/initialized.
          out(i);
       }; // Outputs are 2, 3... 2*a

       for i from 1 to 4 step 2 do { //For loop with step
          out(i);
       }; // Outputs are 1,3

       for i from 4 down to 2 step 2 do { //For loop counting down
          out(i);
       }; // Outputs are 4,2
    }; //end of method meth()
}; //end of struct temp definition.
'>
```

6.3.1.1 C Style For Loop

A C style **for** loop is also available in *e*. The syntax of a C style **for** loop is as follows:

```
for {initial-action; bool-exp; step-action} [do] {action; ...};
```

This **for** loop executes the *initial-action* once, and then checks the *bool-exp*. If the *bool-exp* is TRUE, it executes the *action* block followed by the *step-action*. It repeats this sequence in a loop for as long as *bool-exp* is TRUE. Example 6-8 shows a C style for loop.

Example 6-8 C Style For Loop

```
Example of C style for loop
<'
struct temp {
meth () is {
 var i: int; //Variable i needs to be declared for C style for loop
 var j: int;
 for {i = 0; i < 10; i += 1} do { // C style for loop
   if i % 3 == 0 then {
       continue; //Continue to next iteration
   };
   j = j + i;
   if j > 100 then {
       break; //Break out of the loop
   };
 };
}; //end of method meth()
}; //end of struct temp definition.
'>
```

6.3.2 For each Loop

When iterating through each element of a list, line in a file, or a file in a directory, it is cumbersome to find out the size of the list and then perform the iteration. The **for each** loop solves this problem. The syntax for the **for each** loop is as follows:

```
for each [type] [(item-name)] [using index (index-name)]
        in [reverse] list-exp [do] {action; ...};
```

The **for each** loop executes the *action* block for each item in *list-exp* if its type matches *type*. Inside the *action* block, the implicit variable **it** (or the optional *item-name*) refers to the matched item, and the implicit variable **index** (or the optional *index-name*) reflects the index of the current item. If **reverse** is specified, *list-exp* is traversed in reverse order, from last to first. The implicit variable **index** (or the optional *index-name*) starts at zero for regular loops and is calculated to start at "(list.size() - 1)" for reverse loops. Example 6-9 shows various **for each** loops.

Example 6-9 For each Loops

```
Example of for each loops
<'
extend sys {
    do_it() is {
        var numbers := {8; 16; 24};
        for each in numbers { //for each loop
            print index; //Prints 0,1,2
            print it; //Prints 8, 16, 24
        };

        var sum: int; //Default initial value sum = 0
        for each (n) in numbers { //The for each loop can be applied
                                  //to a list of any type.
            print index; //Prints 0,1,2
            sum += n;
            print sum; //Prints the accumulated total
        };//Value of sum at end of loop is 48

    }; //End of method do_it()
 }; //End of struct extension
'>
```

6.3.3 While Loop

The syntax for the **while** loop is as follows:

```
while bool-exp [do] {action; ...};
```

The **while** loop executes the *action* block repeatedly in a loop while *bool-exp* is TRUE. Example 6-10 shows the usage of **while** loops.

Example 6-10 While Loops

```
Example of while loops
<'
define SMAX 200;
extend sys {
    ctr: uint;
    ctr_assn() is {
        var i: uint;//Default initial value of i = 0
        i = 100; //i is assigned a value 100
        while i <= SMAX { //Loop while value of i <= SMAX
            print ctr;
            ctr = ctr + 10;
            i+=1;
        };
    };
}: //end of sys extension
'>
```

6.3.4 Repeat Loops

The syntax for **repeat** loops is as follows:

```
repeat {action; ...} until bool-exp;
```

A **repeat** loop executes the *action* block repeatedly in a loop until *bool-exp* is TRUE. A **repeat until** action performs the action block at least once. A **while** action might not perform the action block at all. Example 6-11 shows the usage of repeat loops.

Example 6-11 Repeat Loops

```
Example of repeat loops
<'
struct temp {
    i: int;
    meth() is {
      repeat {
          i+=1; //Increment i
          print i; //Print the value of i
      } until i==3; //until the value of i == 3
    }; //end of method meth()
}; //End of struct temp
'>
```

6.4 Useful Output Routines

There are three types of output routines, **out()**, **outf()** and the **do_print()** method. The **out()** routine prints expressions to the output, with a new line at the end. The syntax of the **out()** routine is as follows:

```
out()
out(item: exp, ...);
```

The **outf()** routine prints formatted expressions to output, with no new line at the end. The syntax of the **outf()** routine is as follows:

```
outf(format: string, item: exp, ...);
```

The format string is very similar to C style format syntax with minor variations. The format string for the **outf()** routine uses the following syntax:

```
"%[0|-][#][min_width][.max-chars] (s|d|x|b|o|u)"
```

Table 6-3 describes the characters used in the formatting of the **outf()** string.

Table 6-3 Format String for outf()

0	Pads with 0 instead of blanks.
-	Aligns left. The default is to align right.
min_width	Specifies the minimum number of characters and digits. If there are not enough characters and digits, the expression is padded.
max_chars	Specifies the maximum number of characters and digits. Extra characters and digits are truncated.
s	Converts the expression to a string.
d	Prints a numeric expression in decimal format.
#	Adds 0x before the number. Can be used only with the x (hexadecimal) format specifier. Examples: %#x, %#010x
x	Prints a numeric expression in hex format. With the optional # character, adds 0x before the number.

Table 6-3 Format String for outf() (Continued)

b	Prints a numeric expression in binary format.
o	Prints a numeric expression in octal format.
u	Prints integers (**int** and **uint**) in **uint** format.

Example 6-12 shows the usage of **out()** and **outf()** routines.

Example 6-12 Output Routines

```
Examples of out() and outf() routines
<'
struct pkt {
    protocol: [ethernet, atm, other];
    legal : bool;
    data[2]: list of byte;

};

extend sys {
    pkts[5]: list of pkt; //List of 5 packets
    m1() is { //Method definition
        out(); //Empty new line printed
        out("ID of first packet is   ", pkts[0]); //Print first packet
        //Print formatted data using outf()
        outf("%s %#x","pkts[1].data[0] is   ", pkts[1].data[0]);
        out(); //Empty new line printed
    };
    run()is also {m1()}; //Call m1() from the run method
};
'>
```

The **do_print()** method is a predefined method for any struct or unit and is used to print struct or unit information. The syntax for the **do_print()** method is as follows:

```
[exp.]do_print() ; //exp is either a struct or unit instance.
```

This method is called by the **print** action whenever you print the struct. It is common to invoke the print action which implicitly calls the **do_print()** method for that struct. Example 6-13 shows the invocation of the **print** action.

Example 6-13 Print Action

```
Example shows the usage of a print action.
This print action implicitly calls the do_print()
method for that struct instance. The print action
prints the fields of the struct in a nicely formatted
manner.
<'
struct a {
    i: int;
    s: string;
    // do_print() is a built in method for every struct or unit
};
extend sys {
    a_struct: a; //Instantiate a struct of type a
    m() is {
      print a_struct; //Print the a_struct fields in a nicely
                      //formatted manner. The print action
                      //implicitly calls the a_struct.do_print()
                      //method
    };
};
'>
```

6.5 Summary

- Methods are struct members. *e* methods are similar to C functions, Verilog tasks, and VHDL processes. An *e* method is an operational procedure containing actions that define its behavior.

- A method can have parameters, local variables, and a return value. You can define a method only within a struct and you must create an instance of the struct before you can execute the method.

- Methods can read or write locally declared variables, fields within the local struct, arguments and return value (implicit **result** variable), and fields in other structs using path notation.

- Methods are not executed unless they are invoked. Invoking a method is done by simply calling the method in a predefined or a user-defined method. For example, a method can be invoked from a predefined **post_generate()** method of a **struct**.

- Conditional actions are used to specify code segments that will be executed only if a certain condition is met. The *e* language supports **if-then-else** and **case** actions.

- The four types of iterative actions are **for**, **for each**, **while**, and **repeat** loops.

- There are three types of output routines: **out()**, **outf()**, and the **do_print()** method. The **print** action implicitly calls the **do_print()** method.

6.6 Exercises

1. Define a struct *test1* with two fields, *a1*: 8 bits and *a2*: 8 bits. Define a method *m()* that returns an 8-bit **uint** that is the sum of *a1* and *a2*. Instantiate the struct *test1* in the **sys** struct. Invoke the method *m()* in the **post_generate()** method of the *test1* struct and assign the result to variable *i* using the **out()** method. (Hint: Use the implicit **result** variable.)

2. Define a struct *test2* with two fields, *b1*: 8 bits and *b2*: 8 bits. Define a method *m()* as follows (Hint: Use the **case** action):

   a. Argument *type* can have values of 0, 1, 2 or 3.
   b. Return value is an 8-bit **uint**.
   c. If *type* == 0, then return value = *b1* + *b2*.
   d. If *type* == 1, then return value = *b1* ^ *b2*.
   e. If *type* == 2, then return value = *b1* & *b2*.
   f. If *type* == 3, then return value = *sys.test1.a1* + *sys.test1.a2*.
   g. Print the value of the *type* and **result** using the **outf()** method.
   h. Invoke the method *m()* with *type* == 3 in the **post_generate()** method of the *test2* struct and assign the result to a variable *i*.
   i. Instantiate the struct *test2* in the **sys** struct.

3. Rewrite Exercise 2 using the **if-then-else** action instead of the **case** action. Name the struct *test3* instead of *test2*.

4. Define a struct named "*instruction*" that contains the fields: *opcode*: 8 bits, *op1*: 8 bits and *op2*: 8 bits. Define *l_instructions* as a list of the struct *instruction* in the **sys** struct. In the **sys** struct, define a method *loop()* as follows:

   a. Declare a local variable *count* of the type **uint**.
   b. Declare a local variable *list_size* of the type **uint**.
   c. Set *list_size* equal to the size of the *l_instructions* list.
   d. Write an *e* style **for** loop using the variable *count* that loops from 0 to *list_size-1* in list *l_instructions* and prints out the value of the fields of *instruction* using the **out()** method.
   e. Write a C style **for** loop using the variable *count* that loops from 0 to *list_size-1* in list *l_instructions* and prints out the value of the fields of *instruction* using the **outf()** method.

f. Write a **for each** loop that loops for each element in list *l_instructions* and prints out the value of the struct *instruction* using the **print** action.

g. Invoke the method *loop()* in the **post_generate()** method of the **sys** struct.

5. Extend the **sys** struct. Extend the method *loop()* using the **is also** keywords. Write a **while** loop that counts up each element in list *l_instructions* from 0 to *list_size-1* and prints the struct *instruction* using the **print** action. (Hint: You may need to redeclare the variable *count*.)

6. Load all code from exercises 1 to 5 into Specman Elite and verify that it loads without error.

CHAPTER 7

Events and Temporal Expressions

Timing and synchronization are very important when *e* and HDL processes communicate with each other. The *e* language provides temporal (timing) constructs for specifying and verifying behavior over time. Most *e* temporal language features depend on the occurrence of events, which are used to synchronize activity with a simulator. The temporal language is the basis for capturing the behavior over time for synchronizing with the DUT, protocol checking, and functional coverage. This chapter discusses the *e* temporal language.

Chapter Objectives

- Describe event definition.
- Explain emission of events.
- Understand temporal operators.
- Describe different types of temporal expressions.

7.1 Defining Events

Events define occurrences of certain activity in Specman Elite or the HDL simulator. Events can be attached to temporal expressions (TEs), using the option **is** *temporal-expression* syntax, or they can be unattached. An attached event is emitted when a temporal expression defining it succeeds. The syntax for defining events is as follows:

```
event event-type [is [only] temporal-expression];
```

The components of an event definition are shown in Table 7-1.

Table 7-1 Components of an Event Definition

event-type	The name you give the event type. It can be any legal *e* identifier.
temporal-expression	An event or combination of events and temporal operators.
	To use an event name alone as a temporal expression, you must prefix the event name with the @ sign.

Example 7-1 shows some event definitions. When an event is triggered, it is said to be *emitted*.

Example 7-1 Event Definition

```
Example that shows different types of event definitions.
All e events are triggered (emitted) in Specman Elite.
<'
struct m_str {
    event start_cnt; //This is a standalone event that will be
                     //triggered (emitted) manually when an occurrence
                     //of certain conditions happens.
    event top_clk is fall('~/top/r_clk') @sim;
                    //When a negedge of the HDL signal ~/top/r_clk
                     //occurs, an event top_clk is triggered (emitted)
                    //in Specman Elite. @sim is the sampling event
                     //that initiates a callback from the HDL simulator
                    //to Specman Elite.
    event stop_cnt is {@start_ct; [2]} @top_clk;
                    //The event stop_cnt is emitted when start_cnt
                    //is followed by two cycles of emissions of
                    //top_clk event (falling edge of ~/top/r_clk).
                    //In this case, top_clk
                    //is called the sampling event.
    event clk is rise('~/top/cpu_clk') @sim;
                    //When a posedge of the HDL signal ~/top/cpu_clk
                    //occurs, an event clk is triggered (emitted)
                    //in Specman Elite.
    event sim_ready is change('~/top/ready') @sim;
                    //When a posedge or negedge of the HDL signal
                    //~/top/ready occurs, an event sim_ready is
                    //triggered (emitted) in Specman Elite.
};
'>
```

7.2 Event Emission

The triggering of events is called *event emission*. Events can be emitted explicitly or implicitly. Events are emitted explicitly with the **emit** action. When events are emitted, threads that were blocked, awaiting the emission of this event, are unblocked and scheduled for execution. However, they do not execute until the thread which emitted the event itself blocks. The **emit** event does not consume time. The syntax for the **emit** action is shown below:

```
emit [struct-exp.]event-type;
```

The components of an **emit** action are shown in Table 7-2 below:

Table 7-2 Components of an Emit Action

struct-exp	An expression referring to the struct instance in which the event is defined.
event-type	The type of event to emit.

Example 7-2 shows explicit event emission.

Example 7-2 Explicit Emit Action

```
Example shows how the emit action can be used for
explicit event emission. There is some syntax in this
example that will be covered later in the book. Look
at the comments for the portions that are relevant.
<'
struct xmit_recv { //Define struct xmit_recv
    event rec_ev;
    transmit() @sys.clk is {
        wait cycle; //Wait for the next emission of sys.clk event
        emit rec_ev; //Emit the rec_ev event
        out("rec_ev emitted");
    };
    receive() @sys.clk is {
        wait @rec_ev; //Wait for the next emission of rec_ev event.
                    //This will happen when the rec_ev event is
                    //triggered by the "emit rec_ev action" in the
                    //transmit() method.
        out("rec_ev occurred"); //Print after the rec_ev event has
                                //been emitted.
        stop_run(); //Finish the run
    };
```

Example 7-2 Explicit Emit Action (Continued)

```
    run() is also {
        start transmit(); //Start two processes transmit and
        start receive();  //receive in parallel at the beginning
                          //of simulation.
    };
};

extend sys { //Define struct sys
    event clk is @sys.any; //Create a clk event
    xmtrcv_i: xmit_recv;    //Instantiate the xmit_recv struct
};
'>
```

Events can also be emitted implicitly when they are attached to **@sim** or another temporal expression. There is no need to manually emit these events. 7-3 shows automatic event emission.

Example 7-3 Automatic Event Emission

```
Example shows implicit event emission when various temporal
expressions succeed in Specman Elite or the HDL simulator.
<'
struct m_str {
    event s_event is rise('~/top/start') @sim;
                    //s_event is implicitly triggered when
                    //the ~/top/start HDL signal rises.
                    //@sim is the sampling event
                    //that initiates a callback from the HDL simulator
                    //to Specman Elite.
    event e_event is rise('~/top/end') @sim;
                    //e_event is implicitly triggered when
                    //the ~/top/end HDL signal rises.
    event clk is rise('~/top/clk') @sim;
                    //When a posedge of the HDL signal ~/top/r_clk
                    //occurs, an event top_clk is triggered (emitted)
                    //implicitly in Specman Elite.
    event t_event is {@s_event; [5]; @e_event} @clk;
                    //The event t_event is emitted implicitly
                    //when s_event is followed by 5 posedges of clk
                    //and then an e_event occurs. Thus the t_event
                    //is emitted when a temporal expression succeeds.
};
'>
```

7.3 Event Redefinition

Event definitions can be replaced by means ofthe **is only** keywords. Example 7-4 illustrates event redefinition.

Example 7-4 Event Redefinition

```
Example of event redefinition using the is only keywords
<'
struct m_str {
    event start_ct;
    event top_clk is fall('~/top/r_clk') @sim;
    event stop_ct is {@start_ct; [1]} @top_clk; //Original definition
};
extend m_str {
    event stop_ct is only {@start_ct; [3]}@top_clk; //Event
                                                    //redefinition
};
'>
```

7.4 Sampling Events

Events are used to define the points at which temporal expressions are sampled. An event attached to a temporal expression becomes the sampling event for the temporal expression. The temporal expression is evaluated whenever the sampling event is emitted.

The sampling period is the interval of time from the emission of a sampling event to the next time the sampling event is emitted. All other event emissions within the same sampling period are considered simultaneous. Multiple emissions of a particular event within one sampling period are considered to be one emission of that event.

In Figure 7-1, Q and R are previously defined events that are emitted at the points shown. The temporal expression "$Q@R$" means "evaluate Q every time the sampling event R is emitted." If Q has been emitted since the previous R event, then "$Q@R$" succeeds upon the next emission of R. The final "$Q@R$" success happens because the sampling period for the expression includes the last R event, which is emitted at the same time as the last Q.

Figure 7-1 Sampling Event for a Temporal Expression

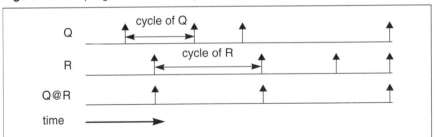

If "Q" in the figure above is a temporal expression that includes other events, "R" is the default sampling event of the temporal expression. The default sampling event of a temporal expression applies to all subexpressions within the expression, except where it is overridden explicitly by another event embedded in the expression.

7.5 Temporal Operators

Events can be converted into basic or complex temporal expressions with temporal operators. Table 7-3 shows a description of the commonly used temporal operators.

Table 7-3 Temporal Operators

Operator Name	Operator Example	Operator Description
Inversion	**not** TE	The **not** temporal expression succeeds if the evaluation of the subexpression does not succeed during the sampling period. Thus **not** TE succeeds on every emission of the sampling event if TE does not succeed.
Failure	**fail** TE	A **fail** succeeds whenever the temporal expression fails. If the temporal expression has multiple interpretations (for example, **fail** (TE1 **or** TE2)), the expression succeeds if and only if all the interpretations fail.
And	TE1 **and** TE2	The temporal **and** succeeds when both temporal expressions start evaluating in the same sampling period and succeed in the same sampling period.

Table 7-3 Temporal Operators (Continued)

Operator Name	Operator Example	Operator Description
Or	TE1 **or** TE2	The **or** temporal expression succeeds when either temporal expression succeeds in the same sampling period.
Sequence	{TE1 ; TE2}	The semicolon (;) sequence operator evaluates a series of temporal expressions over successive emissions of a specified sampling event.
Success Check	**eventually** TE	This operator is used to indicate that the temporal expression should succeed at some unspecified time before the simulation ends.
Fixed Repetition	[*num*]*TE	Repetition of a temporal expression is frequently used to describe cyclic or periodic temporal behavior. The [*num*] fixed repeat operator specifies a fixed number of emissions of the same temporal expression.
First Match Variable Repeat	{[*exp1..exp2*]*TE} @event_name	The first match repeat expression succeeds on the first success of the temporal expression. In this temporal expression, *exp1* is the lower number bound, and *exp2* is the upper number bound.
True Match Variable Repeat	{~[*exp1..exp2*]*TE} @event_name	True match repeat operator is used to specify a variable number of consecutive successes of a temporal expression. *exp1* is the lower number bound and *exp2* is the upper number bound.
Yield	TE1 => TE2	The yield operator is used to assert that success of one temporal expression depends on the success of another temporal expression. The yield expression TE1 => TE2 is equivalent to (**fail** TE1) **or** {TE1 ; TE2}.

Table 7-3 Temporal Operators (Continued)

Operator Name	Operator Example	Operator Description
Delay	**wait delay**(*sim_delay*);	Succeeds after a specified simulation time delay elapses. A callback to Specman Elite occurs after the specified time.
Sample event	TE **@***event_name*	Operator is used to specify the sampling event for a temporal expression. The specified sampling event overrides the default sampling event.
Unary named event	**@***event_name*	An event can be used as the simplest form of a temporal expression. The temporal expression *@event-name* succeeds every time the event occurs. Success of the expression is simultaneous with the emission of the event.
Cycle	**cycle @***event_name*	Represents one cycle of some sampling event.
Boolean Expression	**true**(*exp*) **@***event_name*	The temporal expression succeeds each time the expression *exp* evaluates to TRUE at time of emission of the sampling event.
Edge Expression	**rise/fall/change**(*exp*) **@***event_name*	Detects a change in the sampled value of an expression at a sampling event.

7.6 Temporal Expressions

This section provides a detailed description of different types of temporal expressions and temporal operators.

7.6.1 Basic Temporal Operators

Basic temporal expressions are simple temporal expressions that contain the emission of a particular event. This section includes a discussion of the following:

- Unary named event
- Delay
- Cycle
- Boolean expression
- Edge expression

Example 7-5 shows the usage of some basic temporal expressions.

Example 7-5 Basic Temporal Operators

```
Example shows the usage of basic temporal operators.
<'
struct m_str {\
    //Definition of basic events
    event a_event is rise('~/top/start') @sim;
    event b_event is rise('~/top/end') @sim;
    event clk is rise('~/top/clk') @sim;

    event unary_e is @b_event @clk;
                    //@b_event is a unary event temporal expression.
                    //unary_e occurs at clk event when b_event occurs
                    //in a sampling period.
    event boolean_e is true('~/top/clear' ==1) @clk;
                    //boolean_e is emitted when the boolean
                    //temporal operator checks whether HDL signal
                    //~/top/clear is equal to 1 exactly at the rising
                    //edge of clk.
    event edgep_e is rise('~/top/a' ==1) @clk;
                    //edgep_e is emitted when the edge
                    //temporal operator finds that ~/top/a
                    //has gone from 0 to 1 in the current
                    //sampling period of clk.
    event edgen_e is fall('~/top/a' ==1) @clk;
                    //edgen_e is emitted when the edge
                    //temporal operator finds that ~/top/a
                    //has gone from 1 to 0 in the current
                    //sampling period of clk.
```

Example 7-5 Basic Temporal Operators (Continued)

```
    event edgec_e is change('~/top/a') @clk;
                    //edgec_e is emitted when the edge
                    //temporal operator finds that ~/top/a
                    //has changed value in the
                    //current sampling period of clk.

    temp_oper() @clk is {
      wait delay (10); //Wait for 10 simulation time units
      wait cycle; //Wait for the next emission of clk event
    };
};
'>
```

7.6.2 Sequence Operator

The semicolon (;) sequence operator evaluates a series of temporal expressions over successive emissions of a specified sampling event. Each temporal expression following a ";" starts evaluating in the sampling period following that in which the preceding temporal expression succeeded. The sequence succeeds whenever its final expression succeeds. If any one expression in the list of temporal expressions is missed, the sequence goes back to the beginning and restarts evaluation.

Figure 7-2 shows the results of evaluating the temporal sequence shown below over the series of ev_a, ev_b, and ev_c events shown at the top of the figure. Evaluation of the sequence starts whenever event ev_a occurs.

```
{@ev_a; @ev_b; @ev_c} @qclk;
```

Figure 7-2 Example Evaluations of a Temporal Sequence

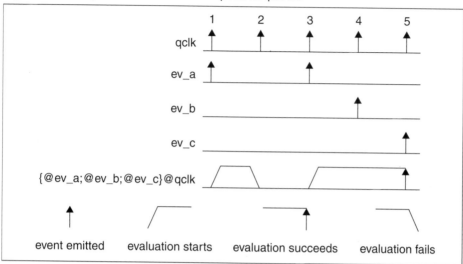

7.6.3 Not and Fail Operators

The **not** temporal expression succeeds if the evaluation of the subexpression does not succeed during the sampling period. Thus **not** TE succeeds on every emission of the sampling event if TE does not succeed.

In Example 7-6, the event *ev_d* occurs every time there is an emission of *ev_c* that is not preceded by an emission of *ev_a* and then two consecutive emissions of *ev_b*.

Example 7-6 Not Operator

```
event ev_d is {not{@ev_a; @ev_b; @ev_b}; @ev_c} @clk;
```

A **fail** succeeds whenever the temporal expression fails. If the temporal expression has multiple interpretations (for example, **fail** (*TE1* **or** *TE2*)), the expression succeeds if and only if all the interpretations fail. The expression **fail** TE succeeds at the point where all possibilities to satisfy TE have been exhausted. Any TE can fail at most once per sampling event.

Example 7-7 shows the usage of a **fail** operator.

Example 7-7 Fail Operator

```
fail {@ev_b;@ev_c}
```

The expression above succeeds for any of the following conditions:

- Event *ev_b* does not occur in the first cycle.
- *ev_b* succeeds in the first cycle, but event *ev_c* is not emitted in the second cycle.

The **fail** operator differs from the **not** operator. Figure 7-3 illustrates the differences in behavior of **not** and **fail** for the sequence of ev_b and ev_c events shown at the top of the figure. To understand the differences, start with the emissions of ev_b and ev_c events shown at the beginning of the figure. Then look at the temporal expression $\{@ev\_b; @ev\_c\}@pclk$ and how that succeeds. Then notice the difference between the application of **not** and **fail** operators to this temporal expression.

Figure 7-3 Comparison of Temporal not and fail Operators

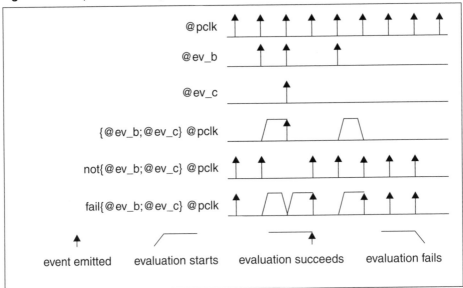

7.6.4 And Operator

The temporal **and** succeeds when both temporal expressions start evaluating in the same sampling period and succeed in the same sampling period. Example 7-8 shows the usage of the **and** operator.

Example 7-8 And Operator

```
event TE3 is (TE1 and TE2) @qclk
```

Evaluation of the above **and** temporal expression for event emissions as shown in Figure 7-4 is as follows:

- Evaluation of both *TE1* and *TE2* begins on the first *qclk*. Both *TE1* and *TE2* succeed between the second and third *qclk* so the event *TE3* is emitted at the third *qclk*.

- The evaluations of *TE1* and *TE2* that begin on the fourth *qclk* eventually result in success of both *TE1* and *TE2*, but *TE2* succeeds before the fifth *qclk*, and *TE1* succeeds before the sixth *qclk*. Therefore, *TE1* **and** *TE2* does not succeed.

- On the seventh *qclk*, evaluation of *TE1* begins, and it succeeds before the ninth *qclk*. However, the corresponding evaluation of *TE2* fails during that period, so *TE3* fails.

Figure 7-4 And Operator

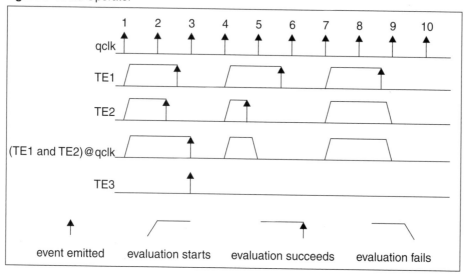

7.6.5 Or Operator

The **or** temporal expression succeeds when either temporal expression succeeds. An **or** operator creates a parallel evaluation for each of its subexpressions. It can create multiple successes for a single temporal expression evaluation. Example 7-9 shows the usage of the **or** operator.

Example 7-9 Or Operator

```
event TE3 is (@TE1 or @TE2) @qclk;
```

Evaluation of the above **or** temporal expression for event emissions as shown in Figure 7-5 is as follows:

- Evaluation of both *TE1* and *TE2* begins on the first *qclk* and succeeds between the second and third *qclk*, so *TE1* **or** *TE2* succeeds at the third *qclk*.

- The evaluations of *TE1* **or** *TE2* that begin on the fourth *qclk* result in success of *TE2* before the fifth *qclk*, so *TE3* succeeds at the fifth *qclk*.

- Evaluation of *TE1* **or** *TE2* begins again at the seventh *qclk*, and *TE1* succeeds before the ninth *qclk*, so *TE3* succeeds at the ninth *qclk*.

Figure 7-5 Example of Temporal or Operator Behavior

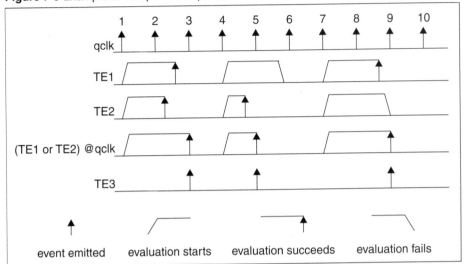

7.6.6 Fixed Repeat Operator

Repetition of a temporal expression is frequently used to describe cyclic or periodic temporal behavior. The fixed repeat operator specifies a fixed number of emissions of the same temporal expression. If the numeric expression evaluates to zero, the temporal expression succeeds immediately.

In Example 7-10, the **wait** action proceeds after the sequence event ev_a, then three emissions of event ev_b, then event ev_c, all sampled at the sampling event $@clk$.

Example 7-10 Fixed Repeat Operator

```
wait {@ev_a; [3]*@ev_b; @ev_c} @clk;
```

7.6.7 First Match Variable Repeat Operator

The first match repeat expression works on multiple emissions of a temporal expression from a lower bound to an upper bound. The first match repeat expression succeeds on the first success of the temporal expression. Example 7-11 shows the usage of the first match variable repeat operator.

Example 7-11 First Match Variable Repeat Operator

```
wait {@ev_a; [0..2]*@ev_b; @ev_c}@pclk;
```

In Example 7-11, the **wait** action proceeds after any one of the three sequences sampled at consecutive sampling events:

- $\{@ev\_a; @ev\_c\} @pclk;$
- $\{@ev\_a; @ev\_b; @ev\_c\}@pclk;$
- $\{@ev\_a; @ev\_b; @ev\_b; @ev\_c\}@pclk;$

In the following example, the **wait** action proceeds after one or more emissions of ev_a at consecutive $pclk$ events, followed by one emission of ev_b at the next $@pclk$ event:

```
wait {[1..]*@ev_a; @ev_b}@pclk;
```

In the following example, the **wait** action proceeds after between zero and three occurrences of the sequence $\{ev\_a; ev\_b\}$ (sampled by $pclk$), followed by an emission of ev_c at the next $pclk$ event:

```
wait {[..3]*{@ev_a; @ev_b}; @ev_c}@pclk;
```

7.6.8 True Match Variable Repeat

The true match repeat expression works on multiple occurrences of a temporal expression from a lower bound to an upper bound. True match variable repeat succeeds every time the subexpression succeeds. This expression creates a number of parallel repeat evaluations within the range.

In Example 7-12, the temporal expression succeeds three $pclk$ cycles after $reset$ occurs, again at four $pclk$ cycles after $reset$, and again five $pclk$ cycles after $reset$ (with $reset$ also sampled at $pclk$):

Example 7-12 True Match Variable Repeat

```
Example shows true match variable repeat operator
<'
struct t {
    event reset;
    event pclk;
    event TE1 is {@reset; ~[3..5]} @pclk; //~ causes true match
                                          //variable repeat operator
};
extend sys {
    t_struct: t;
};
'>
```

The following temporal expression succeeds after any of the sequences $\{A\}$, $\{A; B\}$, $\{A; B; B\}$, or $\{A; B; B; B\}$:

```
{@A;~[..3]*@B}@pclk;
```

7.6.9 Eventually Operator

This operator is used to indicate that the temporal expression should succeed at some unspecified time. Typically, **eventually** is used to specify that a temporal expression is expected to succeed sometime before the simulation ends.

In Example 7-13, the temporal expression succeeds after the event ev_c is followed by event ev_a in the next cycle, and then event ev_b is emitted sometime before the simulation terminates.

Example 7-13 Eventually Operator

```
{@ev_c; @ev_a; eventually @ev_b} @pclk;
```

7.6.10 Yield Operator

The yield operator is used to assert that success of one temporal expression depends on the success of another temporal expression. The yield expression $TE1 => TE2$ is equivalent to (**fail** $TE1$) **or** $\{TE1 ; TE2\}$. The yield expression succeeds without evaluating the second expression if the first expression fails. If the first expression succeeds, then the second expression must succeed in sequence.

In Example 7-14, the temporal expression succeeds if *acknowledge* is emitted 1 to 100 cycles after *request* is emitted. However, if *@request* never happens, then the yield expression succeeds without evaluating the second expression

Example 7-14 Yield Operator

```
Example shows the usage of a yield operator.
<'
struct t {
    event request;
    event acknowledge;
    event clk is rise('~/top/clk') @sim;
    expect @request => {[..99]; @acknowledge} @clk;
    //Note: We would expect @request and @acknowledge
    //to be emitted from a TCM in the same struct or
    //or another struct.
};
'>
```

7.7 Predefined Events

Predefined events are emitted by Specman Elite at particular points in time. The knowledge of these events is useful for debugging and synchronization purposes. Table 7-4 summarizes these predefined events.

Table 7-4 Predefined Events

Predefined Event	Description
sys.any	Emitted on every *e* simulator tick. This is a special event that defines the finest granularity of time. The emission of any event in the system causes an emission of the **any** event at the same tick. For any temporal expression used without an explicit sampling event, **sys.any** is used by default.
sys.tick_start	Emitted at the start of every *e* simulator tick. This event is provided mainly for visualizing and debugging the program flow in the event viewer.
sys.tick_end	Emitted at the end of every *e* simulator tick. This event is provided mainly for visualizing and debugging the program flow in the event viewer.
session.start_of_test	Emitted once at test start. The first action the predefined run() method executes is to emit the **session.start_of_test** event. This event is typically used to anchor temporal expressions to the beginning of a simulation.
session.end_of_test	Emitted once at test end. This event is typically used to sample data at the end of the test. This event cannot be used in temporal expressions as it is emitted after evaluation of temporal expressions has been stopped. The **on session.end_of_test** struct member is typically used to prepare the data sampled at the end of the simulation.
struct.**quit**	Emitted when a struct's **quit()** method is called. Only exists in structs that contain events or have members that consume time (for example, time-consuming methods and **on** struct members). The first action executed during the check test phase is to emit the **quit** event for each struct that contains it. It can be used to cause the evaluation of temporal expressions that contain the **eventually** temporal operator. This allows you to check for **eventually** temporal expressions that have not been satisfied.
sys.new_time	In standalone operation (no simulator), this event is emitted on every **sys.any** event. When a simulator is being used, this event is emitted every time a callback occurs, if the attached simulator's time has changed since the previous callback.

7.8 Summary

- Events define occurrences of certain activity in Specman Elite or the HDL simulator. Events can be attached to temporal expressions (TEs), or they can be unattached. An attached event is emitted automatically when a temporal expression attached to it succeeds.

- The triggering of events is called *event emission*. Events can be emitted explicitly or automatically. Events are emitted explicitly using the **emit** action. The **emit** action does not consume time.
- Event definitions can be replaced using the **is only** keywords.
- Events are used to define the points at which temporal expressions are sampled. An event attached to a temporal expression becomes the sampling event for the temporal expression. The temporal expression is evaluated at every emission of the sampling event.
- Events can be converted into basic or complex temporal expressions with temporal operators. Various temporal operators and their examples were discussed.
- Predefined events are emitted by Specman Elite at particular points in time. The knowledge of these events is useful for debugging and synchronization purposes.

7.9 Exercises

1. Define a struct *test1*. Instantiate the struct *test1* under **sys**. Define the following events in this struct. (Hint: Default sampling event is **@sim**. Although minimal numbers of **@sim** events should be used in *e* code, this exercise uses them for simplicity.)

   a. Event *a_event*.
   b. Event *p_clk* is the rising edge of the HDL signal ~/*top*/*clock*.
   c. Event *n_clk* is the falling edge of the HDL signal ~/*top*/*clock*.
   d. Event *c_clk* is either edge of the HDL signal ~/*top*/*clock*.
   e. Event *rst* is emitted when the HDL signal ~/*top*/*reset* is 1 at sampling event *p_clk* (this is a synchronous reset).

2. Extend the struct *test1*. Add the following events to the struct *test1*.

   a. Event *p_start* is the rising edge of the HDL signal ~/*top*/*packet_valid*.
   b. Event *p_end* is the falling edge of the HDL signal ~/*top*/*packet_valid*.
   c. Event *p_ack* is the rising edge of the HDL signal ~/*top*/*packet_ack*.
   d. Event *p_pulse* is the rising edge of the HDL signal ~/*top*/*packet_pulse*.

3. Extend the struct *test1*. Redefine the following event in struct *test1* using the **is only** syntax.

   a. Event *a_event* is emitted if both *p_start* and *p_pulse* are emitted in the same sampling period. The sampling event is @*p_clk*.

4. Extend the struct *test1*. Define the following events in struct *test1*.

a. Event *b_event* is emitted if either *p_start* or *p_pulse* is emitted in the same sampling period. The sampling event is @*p_clk*.

b. Event *c_event* is emitted if **fail** {*p_start*; *p_ack*} succeeds in a sampling period. The sampling event is @*p_clk*.

c. Event *d_event* is emitted if **not** {*p_start*; *p_ack*} succeeds in a sampling period. The sampling event is @*p_clk*.

d. Event *e_event* is emitted if *p_start* is immediately followed by 25 to 50 clock cycles and then immediately followed by *p_ack*. The sampling event is @*p_clk*.

e. Event *f_event* is emitted if *p_start* is followed eventually by *p_ack*. The sampling event is @*p_clk*.

f. Event *g_event* is emitted if *p_start* is immediately followed by three emissions of *p_pulse*, followed by five clock cycles and then immediately followed by *p_ack*. The sampling event is @*p_clk*.

Time Consuming Methods

In earlier chapters, we discussed methods that execute within a single point of simulation time (within zero time). This type of method is referred to as a regular method. Another type is called a time consuming method or TCM. TCMs execute over multiple cycles. TCMs are used to synchronize processes in an *e* program with processes or events in the DUT. This chapter discusses TCMs.

Chapter Objectives

- Describe time consuming methods (TCMs).
- Explain calling and starting of TCMs.
- Describe **wait** and **sync** actions.
- Understand the **gen** action.
- Explain the use of Verilog tasks and functions from *e* code.
- Describe the use of VHDL procedures and functions from *e* code.

8.1 Defining TCMs

Time consuming methods (TCMs) are *e* methods that are similar to Verilog tasks and VHDL processes. A TCM is an operational procedure containing actions that define its behavior over time. Simulation time elapses in TCMs. TCMs can execute over multiple cycles and are used to synchronize processes in an *e* program with processes or events in the DUT.

TCMs can contain actions that consume time, such as **wait** and **sync**, and can call other TCMs. Within a single *e* program, multiple TCMs can execute either in sequence or in parallel, along separate threads. A TCM can also have internal branches, which are multiple action blocks executing concurrently. A TCM can have parameters, local variables, and a return value.

A TCM can be defined only within a struct or unit and an instance of the struct or unit must be created before you can execute the TCM. When a TCM is executed, it can manipulate the fields of that struct instance. The syntax for a TCM is shown below:

```
method-name ([parameter-list]) [: return-type]@event is {action;…};
```

Table 8-1 describes the components of a TCM definition.

Table 8-1 Components of a TCM Definition

method-name	A legal *e* name.
parameter-list	A list composed of zero or more parameter declarations of the form *param-name*: [*]*param-type* separated by commas. The parentheses around the parameter list are required even if the parameter list is empty.
	param-name A legal *e* name.
	* When an asterisk is prefixed to a scalar parameter type, the location of the parameter, not its value, is passed. When an asterisk is prefixed to a list or struct type, the method can completely replace the struct or list.
	param-type Specifies the parameter type.
return-type	For methods that return values, specifies the data type of the return value.
@event	Specifies a default sampling event that determines the sampling points of the TCM. This event must be a defined event in *e* and serves as the default sampling event for the TCM itself as well as for time consuming actions, such as **wait**, within the TCM body. Other sampling points can also be added within the TCM.
action;…	A list of zero or more actions, either time consuming actions or regular actions.

Example 8-1 illustrates a TCM definition.

Example 8-1 TCM Definition

```
Example shows a simple TCM definition
<'
struct meth {
    event pclk is rise('~/top/pclk')@sim; //Define event
    event ready is rise('~/top/ready')@sim; //Define event
    event init_complete; //Define event
    init_dut() is empty; //Define empty method (regular method)

    my_tcm() @pclk is { //Define a TCM called my_tcm(). It becomes
                        //a TCM by virtue of the @pclk sampling event.
                        //A TCM can have constructs that
                        //elapse simulation time. @pclk becomes the
                        //default sampling event of any timing
                        //construct.
        wait @ready; //Wait until the ~/top/ready HDL signal rises
        wait [2]; //Wait for two occurrences of @pclk (default
                  //sampling event of the TCM)
        wait [3] @ready; //Wait for three occurrences of @ready
                         //(Override default sampling event of the
                         //TCM which is @pclk)
        init_dut(); //Call a regular method
        emit init_complete; //Manually trigger the init_complete event
    }; //end of my_tcm() definition
};
'>
```

8.1.1 Characteristics of TCMs

The following restrictions apply to all TCMs:

- Similar to a regular method, a TCM can read or write locally declared variables or fields within the local struct, accept arguments and return a value (implicit **result** variable), and use fields in other structs using path notation.

- The maximum number of parameters you can declare for a TCM is 14. Since a TCM can access all fields in the verification hierarchy using the hierarchical path notation, very few values are passed as arguments to the TCM. Therefore the limit of 14 arguments is not a major limitation. Moreover, it is always possible to work around this restriction by passing a compound parameter such as a struct or a list.

- TCMs cannot have variable argument lists. It is possible to work around this restriction by passing a list, which can have variable lengths, or a struct, which can have conditional fields.

8.2   Invoking TCMs

Before invoking a TCM, you must create an instance of the struct that contains it. TCMs can be called or started. Table 8-2 shows the rules for calling or starting TCMs and methods (regular methods).

Table 8-2 Rules for Calling and Starting TCMs and Methods

Calling Procedure	Can Call	Can Start
Method	Another Method	A TCM
TCM	Another TCM Another Method	Another TCM

8.2.1   Calling TCMs

You can execute a TCM by *calling* it from another TCM. The TCM that is called may or may not return a value. A call of a TCM that does not return a value is syntactically an action.

A called TCM begins execution either when its sampling event is emitted or immediately, if the sampling event has already been emitted for the current simulator callback. The calling TCM is blocked until the called TCM returns before continuing execution. For this reason, a called TCM is considered a subthread of the calling TCM and shares the same thread handle (thread ID) with the calling TCM.

Figure 8-1 shows the sequence for executing a *called* TCM (TCM2). The calling TCM (TCM1) must wait for TCM2 to complete before continuing execution. Therefore, TCM1 *blocks* until TCM2 completes execution.

Figure 8-1 Calling TCMs

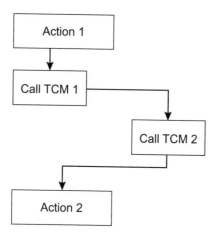

Example 8-2 shows the calling of a TCM.

Example 8-2 Calling a TCM

```
Example shows how to call a TCM from another TCM.
In this example init_dut() TCM is called from the my_tcm()
TCM.
<'
struct meth {
    event pclk is rise('~/top/pclk')@sim; //Define event
    event ready is rise('~/top/ready')@sim; //Define event
    event init_complete; //Define event
    init_dut() @pclk is { //Define TCM init_dut. @pclk is the
                          //default sampling event.
        wait [5] @ready; //Wait for five occurrences of @ready signal
        wait [1]; //Wait for one cycle of @pclk signal
        out("Finishing the init_dut TCM");
    };
```

Example 8-2 Calling a TCM (Continued)

```
    my_tcm() @pclk is { //Define a TCM called my_tcm().
                        //@pclk is the default sampling event.
        wait @ready; //Wait until the ~/top/ready HDL signal rises
        wait [2]; //Wait for two occurrences of @pclk (default
                //sampling event of the TCM)
        wait [3] @ready; //Wait for three occurrences of @ready
                        //(Override default sampling event of the
                        //TCM that is @pclk)
        init_dut(); //Call the init_dut() TCM. Wait until init_dut()
                //finishes.
        emit init_complete; //Manually trigger the init_complete event
    }; //end of my_tcm() definition
};
'>
```

8.2.2 Starting a TCM

You can execute a TCM in parallel with other TCMs by *starting* it from another method or TCM using the keyword **start**. A TCM that is started may not be a value returning TCM.

A started TCM begins execution either when its sampling event is emitted or immediately, if the sampling event has already been emitted for the current simulator callback.

A started TCM runs in parallel with the TCM that started it in a separate thread. A started TCM has a unique thread handle (thread ID) that is assigned to it automatically by the Specman Elite scheduler. The recommended way is to start an initial TCM by extending the related struct's pre-defined **run()** method. This initial TCM can then call other TCMs.

Figure 8-2 shows the sequence for executing a *started* TCM (TCM2). The starting TCM (TCM1) does not wait for TCM2 to complete before continuing execution. TCM1 simply spawns TCM2 in zero simulation time and continues execution.

Figure 8-2 Starting TCMs

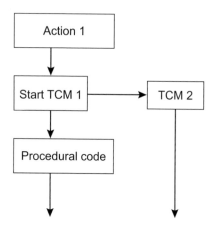

Example 8-3 shows the starting of a TCM.

Example 8-3 Starting a TCM

```
Example shows how to start a TCM from another method or TCM.
In this example my_tcm() TCM is started from the run()
method of the struct.
<'
struct meth {
    event pclk is rise('~/top/pclk')@sim; //Define event
    event ready is rise('~/top/ready')@sim; //Define event
    event init_complete; //Define event
    my_tcm() @pclk is { //Define a TCM called my_tcm()
                        //@pclk is the default sampling event
        wait @ready; //Wait until the ~/top/ready HDL signal rises
        wait [2]; //Wait for two occurrences of @pclk (default
                //sampling event of the TCM)
        wait [3] @ready; //Wait for three occurrences of @ready
                        //(Override default sampling event of the
                        //TCM which is @pclk)
        emit init_complete; //Manually trigger the init_complete event
    }; //end of my_tcm() definition
```

Example 8-3 Starting a TCM (Continued)

```
    run() is also { //Predefined run() method in every struct
                    //Needs to be extended to start my_tcm()
        out("Starting my_tcm..."); //Print message
        start my_tcm(); //my_tcm() is spawned off as a new thread
                        //at simulation time 0 from the run() method
        };
};
'>
```

8.2.3 Extending a TCM

A TCM can be extended in the same way as a method by means of the **is also**, **is first**, and **is only** keywords. Example 8-4 illustrates a TCM extension.

Example 8-4 Extending a TCM

```
Example shows how to extend a TCM.
Can use the following syntax
is also - Adds to the end of existing TCM code.
is first - Adds to the beginning of existing TCM code.
is only - Replaces existing TCM code.
<'
type ctrl_cmd_kind: [RD, WR];

struct ctrl_cmd {
    kind: ctrl_cmd_kind;
    addr: int;
};
```

Example 8-4 Extending a TCM (Continued)

```
struct ctrl_stub {
    init_commands: list of ctrl_cmd;
    event cclk is rise('top.control_clk') @sim;

    init_dut() @cclk is { //TCM
        for each (cmd) in init_commands {
            execute(cmd);
        };
    };

    execute(cmd: ctrl_cmd) @cclk is { //Original definition of TCM
        out(appendf("Executing a %s (addr %s) control command",
            cmd.kind, cmd.addr));
        case cmd.kind {
            RD: {
                wait [2];
            };
            WR: {
                wait [3];
            };
        };
    };
};

extend ctrl_stub {
    execute(cmd: ctrl_cmd) @cclk is first { //Extension of TCM
        if ('top.interrupt' == 1) then { //Add code to the beginning
                                         //of existing TCM code
            return;
        };
    };
};
'>
```

8.3 Wait and Sync Actions

There are two actions that are used to synchronize temporal test activities within *e* and between the DUT and *e* simulator. These actions are **wait** and **sync**. These actions are called within TCMs. This section discusses these actions.

8.3.1 Wait Action

The **wait** action suspends the execution of the current TCM until a given temporal expression succeeds. If no temporal expression is provided, the TCM waits for its default sampling event. The syntax for the **wait** action is shown below. The **until** keyword is purely optional and has no effect on the behavior of the **wait** action. The **wait** action suspends the execution of the TCM until the temporal expression succeeds.

```
wait [[until] temporal-expression ];
```

A TCM cannot continue during the same cycle in which it reaches a **wait**. Even if the temporal expression succeeds in the current simulator callback, the TCM will *not* proceed. The TCM has to wait at least until the next simulator callback to evaluate the temporal expression and check for success or failure. Therefore, the **wait** action always requires at least one cycle of the TCMs sampling event before execution can continue. Example 8-5 shows numerous variations of the **wait** action.

Example 8-5 Wait Action

```
Example shows variations of the wait action.
A wait action is called only inside a TCM.
<'
struct wt {
    var1: uint (bits: 4);
    var2: uint (bits: 4);
    go_wait()@sys.clk is {
        wait [3]*cycle;
        // Continue on the fourth cycle from now
        wait delay(30);
        // Wait 30 simulator time units
        wait [var1 + var2]*cycle;
        // Calculate the number of cycles to wait

        wait until [var1 + var2]*cycle;
        // Same as wait [var1 + var2]*cycle
        wait true(sys.time >= 200);
        // Continue when sys.time is greater than or equal to 200
        wait cycle @sys.reset;
        // Continue on reset even if it is not synchronous with
        // the TCMs default sampling event
        wait @sys.reset;
        // Continue on the next default sampling event after reset
    }; //End of TCM go_wait()
        run() is also { //Start the go_wait TCM from the run() method.
        start go_wait();
    };
};
```

Example 8-5 Wait Action (Continued)

```
extend sys {
    wt_i: wt;
    event clk;
    event reset;
};
'>
```

8.3.2 Sync Action

The **sync** action suspends execution of the current TCM until the temporal expression succeeds. Evaluation of the temporal expression starts immediately when the **sync** action is reached. If the temporal expression succeeds within the current simulator callback, the execution continues immediately. The syntax for the **sync** action is shown below.

```
sync [temporal-expression];
```

The TCM suspends until the emission of its sampling event, or continues immediately if the sampling event has been emitted in the current simulator callback. The **sync** action is similar to the **wait** action, except that a **wait** action always requires at least one cycle of the TCMs sampling event before execution can continue. With a **sync** action, execution can continue in the same simulator callback. Example 8-6 shows one variation of the **sync** action.

Example 8-6 Sync Action

```
Example shows the sync action.
A sync action is called only inside a TCM.
Differences between wait and sync are also
shown.
<'
struct data_drive {
    event clk is rise('top.clk') @sim;
    data: list of int;
    driver() @clk is {
        for each in data {
            wait true('~/top/data_ready'== 1);
                // Will not fall through, even if the condition
                // holds when the wait is evaluated.
                '~/top/in_reg' = it;
        };
        stop_run(); //Finish the simulation run.
    };
```

Example 8-6 Sync Action (Continued)

```
    shadow() @clk is {
        while TRUE {
            sync true('~/top/data_ready' == 0);
                // If the condition holds, the sync falls through
                // as soon as the sync is evaluated.
            out("Shadow read ", '~/top/in_reg');
            wait cycle;
                // This wait is necessary to prevent
                // a zero time loop.
            //Even though this is a while TRUE (infinite) loop,
            //the stop_run() call in driver() TCM will ensure
            //that the simulation finishes.
        };
    };
    run() is also { //The run method must be extended.
        start driver(); // Spawn the driver TCM thread
        start shadow(); // Spawn the shadow TCM thread.
    };
};
'>
```

8.3.3 Difference between Wait and Sync Actions

The **sync** action is similar to the **wait** action, except that a **wait** action always requires at least one cycle of the TCMs sampling event before execution can continue. With a **sync** action, execution can continue in the same simulator callback. Figure 8-3 shows the fundamental difference in the usage of **wait** and **sync** actions.

Figure 8-3 Usage of Wait and Sync Actions

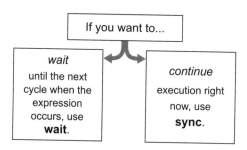

Figure 8-4 explains the difference between **wait** and **sync** further. Only a portion of the code from a TCM block is shown in this example. The *print a;* action happens exactly at time *t0*. The wait and sync actions are both evaluated at time t0. At this time, the temporal expression *true('~/ top/enable' == 1)@clk* evaluates successfully. Therefore, the **sync** action falls through right away and *print c;* action also happens at time *t0*. However, the **wait** action still must wait until the next success of *true('~/top/enable' == 1)@clk* which happens only at time *t1*. Therefore, the *print b;* action is executed at time *t1*.

Figure 8-4 Wait and Sync Actions (Scenario 1)

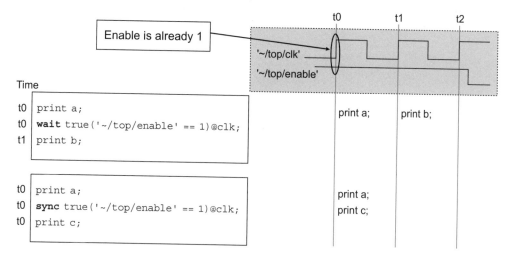

Figure 8-5 shows a scenario in which **wait** and **sync** behave identically. The *print a;* action happens exactly at time *t0*. The wait and sync actions are both evaluated at time t0. At this time, the temporal expression $true('\sim/top/enable' == 1)@clk$ does not evaluate successfully. Therefore, both **wait** and **sync** must wait until the next success of $true('\sim/top/enable' == 1)@clk$ which happens only at time $t1$. Therefore, both *print b;* and *print c;* actions are executed at time $t1$.

Figure 8-5 Wait and Sync Actions (Scenario 2)

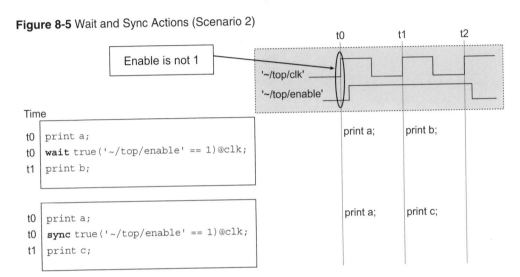

8.3.4 Implicit Sync at the Beginning of Every TCM

There is always an implicit **sync** on the sampling event at the beginning of every TCM. This means that even though the TCM is called or started, it begins execution immediately only if the sampling event has been emitted in the current simulator callback. Otherwise, the TCM suspends execution until the first emission of the sampling event. Figure 8-6 shows how an implicit sync is present at the beginning of every TCM.

Figure 8-6 Implicit Sync at the Beginning of Every TCM

```
<'
struct port {
    reset_delay: uint;

    event clk is rise('top.rclk');

    reset()@clk is {
                        sync @clk;
        'top.reset' = 0b1;
        wait [reset_delay] * cycle;
        'top.reset' = 0b0;
        wait [1] * cycle;
    };
    run() is also { start reset();};
};
'>
```

Sampling **event** @clk must occur before TCM execution begins. This causes an *implicit* **sync** action.

8.4 Gen Action

One can generate the fields of a struct on-the-fly by calling the **gen** action from a TCM or a method.[1] The **gen** action generates a random value for the instance of the item specified in the expression and stores the value in that instance, while considering all relevant constraints at the current scope on that item or its children. Constraints defined in a scope above the scope from which the **gen** action is executed are ignored. It is possible to generate values for particular struct instances, fields, or variables during simulation (on-the-fly generation) with the **gen** action. It is not possible to use the **gen** action to generate units. The syntax for the **gen** action is as follows:

```
gen gen-item [keeping {[it].constraint-bool-exp; …}];
```

Additional constraints can be applied to a particular instance of the item by use of the **keeping** keyword. This constraint applies for that instance in addition to all other constraints applicable to the item from its original **struct** definition. The fields of the **gen** action are shown below in Table 8-3.

Table 8-3 Fields of Gen Action

gen-item	A generatable item. If the expression is a struct, it is automatically allocated, and all fields under it are generated recursively, in depth-first order.

---

1. Although the **gen** action is introduced in the chapter on TCMs, it can also be called from non-TCM methods or constructs.

Table 8-3 Fields of Gen Action (Continued)

constraint-bool-exp	A simple or a compound boolean expression.

Example 8-7 uses the **gen** action within a TCM called *gen_next( )* to create packets to send to the DUT. A constraint within the **gen** action keeps *len* with a range of values. A constraint defined at a lower scope level, *packet*, is also applied when the **gen** action is executed, keeping the size of the *data* field equal to the *len* field. The constraint defined at the **sys** level **keep** *sp.next_packet.len* == 4; is not considered because it is not at the current scope of the **gen** action.

Example 8-7 Gen Action

```
Example shows the use of a gen action.
<'
extend sys {
    event event_clk is rise('~/top/clk')@sim; //define clock event
    sp: send_packet; //Instantiate send_packet
    keep sp.next_packet.len == 4; //constraint on sp.next_packet.len
};

struct packet { //Define packet struct
    len: int [0..10];
    kind: [normal, control, ack];
    data: list of int;
    keep me.data.size() == len;
};
```

Example 8-7 Gen Action (Continued)

```
struct send_packet { //Struct to send the packet.
    num_of_packets_to_send: int [0..20]; //Series of packets
    !next_packet: packet; //A packet instance with do-not-generate

    gen_next() @sys.event_clk is { //Define TCM
        gen num_of_packets_to_send; // Random loop delimiter
        for i from 0 to num_of_packets_to_send - 1 do {
            wait ([100]*cycle);
            gen next_packet keeping { // gen the packet
                .len in [5..10]; //additional constraint on len
                .kind in [normal, control]; //constraint on kind
                //This will generate a packet with len in 5..10 instead
                //of 0..10 and kind will be either normal or control
                //but not ack.
            }; //end of gen action
        }; //end of for loop
    }; //end of gen_next() TCM
    run() is also {
        start gen_next(); //Start the TCM at time 0.
    };
};
'>
```

8.5   Using HDL Tasks and Functions

In TCMs, it is possible to call HDL tasks and functions directly from *e* code. This is useful when there are existing bus functional models (BFMs) in HDL that are already proven to work. Figure 8-7 shows such a scenario.

Figure 8-7 Use of HDL Task and Function Calls in *e*

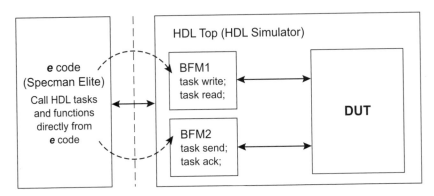

8.5.1 Verilog Task

A Verilog task definition is a statement or a unit member. The syntax for a Verilog user-defined task or system task in *e* so that it can be called from a TCM is as follows:

```
verilog task 'HDL-pathname' (verilog-task-parameter[, …]);
```

Example 8-8 shows the usage of a Verilog task call from *e* code.

Example 8-8 Verilog Task

```
Example shows the usage of a Verilog task call from
e code.
<'
struct mem_w {
    addr: int;
    data: int(bits: 64);
};
```

Example 8-8 Verilog Task (Continued)

```
unit receiver {
    verilog task 'top.write_mem' //Actual task in Verilog top module
      (addr:32:in,data:64:out,status:32:out);
                              //Arguments of the above Verilog
                              //task. 32, 64 etc are arg widths
    event mem_read_enable; //Event definition.

    get_mem(mw: mem_w) @mem_read_enable is { //Define TCM
        var error_status: int; //Local variable

        //Call the write_mem task defined in Verilog 'top' module
        'top.write_mem'(mw.addr, mw.data, error_status);
        check that error_status == 0; //
    };
};
'>
```

8.5.2 Verilog Function

A Verilog function definition is a statement or a unit member. The syntax for a Verilog user-defined function or system task in *e* so that it can be called from a TCM is as follows:

```
verilog function 'HDL-pathname' (verilog-function-parameter[, … ]):
result-size-exp ;
```

Example 8-9 shows the usage of a Verilog function call from *e* code.

Example 8-9 Verilog Function

```
Example shows the usage of a Verilog function call from
e code.
<'
//Verilog function statement, defined outside of unit
//Two arguments addr (32 bits), data (64 bits).
//Return value of function is 32 bits wide.
verilog function 'top.write_mem'(addr:32,data:64):32;
```

Example 8-9 Verilog Function (Continued)

```
unit memory_driver {
    event mem_enable is rise ('top.enable') @sim;
    write() @mem_enable is {
        var error_status: int;
        error_status = 'top.write_mem'(31,45); //Call function
    };
};
'>
```

8.5.3 VHDL Procedure

A VHDL procedure definition is a statement or a unit member. The syntax for a VHDL procedure or system task in *e* so that it can be called from a TCM is as follows:

```
vhdl procedure 'identifier' using option, … ;
```

Example 8-10 shows the usage of a VHDL procedure call from *e* code.

Example 8-10 VHDL Procedure

```
Example shows the usage of a VHDL procedure call from
e code.
<'
unit transactor {
    vhdl procedure 'send_packet' using library="work",
        package="pkg";

    test() @sys.clk is {
        'work.pkg.send_packet'(); //Call VHDL procedure
    };
};

extend sys {
    event clk is rise ('top.clk') @sim;

    transactor1: transactor is instance;
    transactor2: transactor is instance;

    run() is also {
        start transactor1.test();
        start transactor2.test();
    };
};
'>
```

8.5.4 VHDL Function

A VHDL function definition is a statement or a unit member. The syntax for a VHDL procedure or system task in *e* so that it can be called from a TCM is as follows:

```
vhdl function 'identifier' using option, …;
```

Example 8-11 shows the usage of a VHDL function call from *e* code.

Example 8-11 VHDL Function

```
Example shows the usage of a VHDL function call from
e code.
<'
//VHDL function statement
vhdl function 'increment' using
    interface="(a: integer) return integer",
    library="work", package="pkg", alias="integer_inc_1";

//VHDL function statement
vhdl function 'increment' using
    interface="(a: integer; n: integer) return integer",
    library="work", package="pkg",
    alias="integer_inc_n";

extend sys {
    event clk is rise ('top.clk') @sim;

    test(a:int)@clk is { //Define a TCM
        //Make calls to VHDL functions
        check that 'integer_inc_1'(a) == 'integer_inc_n'(a,1);
    };
};
'>
```

8.6 Summary

- Time consuming methods (TCMs) are *e* methods that are similar to Verilog tasks and VHDL processes. A TCM is an operational procedure containing actions that define its behavior over time. Simulation time elapses in TCMs. TCMs can execute over multiple cycles and are used to synchronize processes in an *e* program with processes or events in the DUT.

- TCMs can contain actions that consume time, such as **wait**, and **sync**, and can call other TCMs. Within a single *e* program, multiple TCMs can execute either in sequence or in parallel, along separate threads.

- A TCM can be defined only within a struct or unit and an instance of the struct or unit must be created before you can execute the TCM. When a TCM is executed, it can manipulate the fields of that struct instance.

- Similar to a regular method, a TCM can read or write locally declared variables, fields within the local struct, arguments and return value (implicit **result** variable), and fields in other structs using path notation.

- The maximum number of parameters you can declare for a TCM is 14. However, it is always possible to work around this restriction by passing a compound parameter such as a struct or a list.

- The **wait** action suspends the execution of the current time consuming method until a given temporal expression succeeds. A TCM cannot continue during the same cycle in which it reaches a **wait**.

- The **sync** action suspends execution of the current TCM until the temporal expression succeeds. Evaluation of the temporal expression starts immediately when the **sync** action is reached.

- The **sync** action is similar to the **wait** action, except that a **wait** action always requires at least one cycle of the TCMs sampling event before execution can continue. With a **sync** action, execution can continue in the same simulator callback.

- There is always an implicit **sync** on the sampling event at the beginning of every TCM.

- One can generate the fields of a struct on-the-fly by calling the **gen** action from a TCM. The **gen** action generates a random value for the instance of the item specified in the expression and stores the value in that instance, while considering all relevant constraints at the current scope on that item or its children. Additional constraints can be applied to a particular instance of the item by use of the **keeping** keyword.

- In TCMs, it is possible to call HDL tasks and functions directly from *e* code. This is useful when there are existing bus functional models (BFMs) in HDL that are already proven to work. *e* supports Verilog tasks and functions and VHDL procedures and functions.

8.7 Exercises

1. Define a unit *test1*. Define the following elements in this unit.

    a. Event *p_clk* is the rising edge of the HDL signal ~/*top*/*clock*.

    b. Event *ready* is the rising edge of the HDL signal ~/*top*/*ready*.

    c. Define TCM *ready_cycle()* with @*p_clk* as the sampling event.

    d. First TCM action is to wait for 5 cycles of @*p_clk*.

    e. Second TCM action is to wait for 6 cycles of @*ready* @*p_clk*.

    f. Third TCM action is to wait for 10 cycles of @*p_clk*.

    g. Fourth TCM action is to print "Ready cycle complete" using the **out**() action.

  h. End the TCM.

  i. Extend the **run**() method of *test1*.

  j. Inside the **run**() method, **start** *ready_cycle()*.

  k. Extend the struct **sys**.

  l. Instantiate the unit *test1* under **sys**.

2. Extend the unit *test1*. Define the following elements in this unit.

  a. Extend TCM *ready_cycle()*.

  b. Add a print "Ready cycle start" to the beginning of the TCM using the **out**() action.

  c. Add a sync *@ready* after this print.

3. Create a struct *data*. Define the following fields in this struct.

  a. *address* (8 bits).

  b. Constraint on *address* to be in range 10..20.

  c. *data* (16 bits).

  d. Constraint on *data* to be in range 120..200.

4. Extend the unit *test1*. Define the following elements in unit *test1*.

  a. Instantiate the struct *data*. The instance name is *data1*. Mark it as a do-not-generate field.

  b. Extend TCM *ready_cycle()*.

  c. Declare a local variable *count*.

  d. Generate value for *count* using the **gen** action.

  e. Write a for loop from 0 to *count-1*.

  f. In the loop, generate a value for *data1* using the **gen** action.

  g. Add additional constraints for generation of *data1*, *address* to be in range 15..17 and *data* to be in range 150..180.

  h. Print the values of the fields of *data1* using the **print** action.

  i. End the for loop.

  j. End the TCM extension.

  k. End the extension for unit *test1*.

5. Explain how to call Verilog tasks and functions from *e* code.

6. Explain how to call VHDL procedures and functions from *e* code.

CHAPTER 9

Checking

Automatic checking is a very important aspect of verification with e. There are two types of checks, data checks and temporal checks. For a good methodology, it is recommended that checks must also be separated from the drivers and test scenarios. The e language provides many constructs to facilitate checking. This chapter discusses the e constructs to perform checks.

Chapter Objectives

- Describe **pack()** and **unpack()** methods.
- Describe physical and virtual fields.
- Describe comparison of two structs.
- Explain the **check that** construct.
- Describe **dut_error()** method.
- Understand how to set check failure effects.
- Explain temporal checking with the **expect** and **on** struct members.

9.1 Packing and Unpacking

A DUT requires data on its input HDL signals. You can drive scalar fields of structs directly on to HDL signals. However, a lot of the verification engineer's time is often spent in tracking the bit fields for a specific protocol.

An ideal scenario is one in which the verification engineer works with abstract structs at the input and the output and there is an automatic way to convert these abstract struct instances to a list of bits and bytes. These converted bits or bytes are then applied to the HDL signals of the DUT input.

The *e* language provides the **pack()** and **unpack()** mechanism to do this conversion. However, in order to understand these conversions, it is necessary to introduce the concept of *physical* fields.

9.1.1 Physical Fields

There are two types of fields in any struct, *physical* fields and *virtual* fields. Physical fields have the following characteristics:

- They are typically extracted from the design specification.
- Their values need to be injected into the DUT.

Virtual fields have the following characteristics:

- These fields have no direct meaning in the design specification.
- These fields improve controllability and generation of the fields of the struct.
- These fields define variations of data items (**when** subtyping).
- These fields are required for checking, debugging etc.

The *e* language syntax provides a way of tagging physical fields by preceding these fields with a % symbol. Physical fields are significant only for the **pack()** and the **unpack()** method. Otherwise, their behavior is identical to virtual fields. In other words, the difference between physical and virtual fields is limited to their behavior with the **pack()** and the **unpack()** method. Example 9-1 shows the definition of physical fields.

Example 9-1 Physical Fields

```
Example that shows the difference between virtual fields and
physical fields. Anything that is not tagged with a % symbol is
a virtual field.
<'
struct my_packet{
    atm: bool; //Virtual field
    type: [short, medium, long]; //Virtual field
    %addr:  int(bits:4); //Physical field
    %len: int(bits:10); //Physical field
    init() is also {
        addr = 3;
        len = 15;
    };
    keep type == short => len in [5..10]; //Virtual fields are used
                                          //to control generation
    when TRUE'atm my_packet { //Virtual fields are used for subtypes
        %trailer: int(bits:8); //Add an extra physical field
    };
};
'>
```

9.1.2 Packing

Packing[1] is commonly used to prepare high-level *e* data into a form that can be applied to a DUT. Packing performs concatenation of items, including items in a list or fields in a struct, in the order specified by the pack options parameter and returns a list of bits, bytes, or any uint/int size. The return value is automatically cast to its assigned item. This method also performs type conversion between any of the following:

- scalars
- strings
- lists and list subtypes (same type but different width)

The syntax for the **pack()** method is as follows:

```
pack(option:pack option, item: exp, ...): list of bit;
```

Table 9-1 shows the components of a **pack()** method call.

Table 9-1 Components of pack() Method

option	For basic packing, this parameter is one of the following:
packing.high	Places the least significant bit of the last physical field declared or the highest list item at index [0] in the resulting list of bit. The most significant bit of the first physical field or lowest list item is placed at the highest index in the resulting list of bit.
packing.low	Places the least significant bit of the first physical field declared or lowest list item at index [0] in the resulting list of bit. The most significant bit of the last physical field or highest list item is placed at the highest index in the resulting list of bit.
NULL	If NULL is specified, the global default is used. This global default is set initially to **packing.low**.
item	A legal *e* expression that is a path to a scalar or a compound data item, such as a struct, field, list, or variable.

---

1. The **pack()** method is used more frequently in driving stimulus than in checking. However, this is the most relevant place to introduce this concept.

Example 9-2 shows the usage of the **pack()** method.

Example 9-2 Usage of pack() Method

```
Example shows the use of the pack() method
to pack an entire struct or a list of fields.
When packing an entire struct, the physical
fields are picked up in the order in which they
are defined. Packing a list of fields performs
a concatenation of those fields.

There are two options most commonly used with
the pack() method:
- packing.high: Starts filling the fields into higher
order positions.
- packing.low: Starts filling the fields into lower
order positions.
<'
struct packet {
    %dest        : int (bits : 8); //Physical Field
        keep dest == 0x55; //Constraint
    %version    : int (bits : 2); //Physical Field
        keep version == 0x0; //Constraint
    %type_pkt      : uint (bits : 6); //Physical Field
        keep type_pkt == 0x3f; //Constraint
    %payload : list of int(bits:4); //Physical Field
        keep payload.size() == 2; //Constraint
        keep for each in payload { //Constraint
            it == index;
          };

    //Fields to hold the packed data
    //We could also declare these as list of byte
    !data_packed_low : list of bit;
    !data_packed_high : list of bit;
    !data_packed_fields : list of bit;

    post_generate() is also { //Call the pack() method here
       //Entire struct packet is packed with packing.low
        data_packed_low = pack(packing.low, me);//Using packing.low
        //Entire struct packet is packed with packing.high
        data_packed_high = pack(packing.high, me);//Using packing.high
        //Individual fields of the packet are packed with packing.low
        data_packed_order = pack(packing.low, dest, version, type,
                                payload);//Using packing.low
    };
};
'>
```

Figure 9-1 shows the order in which the fields are packed for the three possible scenarios listed above. The resulting list of bits is 24 bits wide.

Figure 9-1 Order of Packing

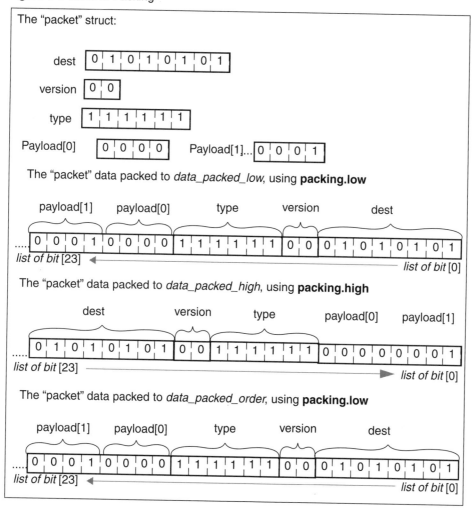

9.1.3 Unpacking

The **unpack**() method does exactly the opposite of packing. Unpacking is commonly used to convert a raw bit or byte stream into high level data by storing the bits of the value expression into the target expressions. Unpacking operates on scalar or compound (struct, list) data items. This is useful for data checking. The syntax for the **unpack()** method is as follows:

```
unpack(option: pack option, value: exp, target1: exp [, target2: exp,
...];
```

Table 9-2 shows the components of an **unpack()** method call.

Table 9-2 Components of unpack() Method

option	For basic packing, this parameter is one of the following.
packing.high	Places the most significant bit of the list of bit at the most significant bit of the first field or lowest list item. The least significant bit of the list of bit is placed into the least significant bit of the last field or highest list item.
packing.low	Places the least significant bit of the list of bit into the least significant bit of the first field or lowest list item. The most significant bit of the list of bit is placed at the most significant bit of the last field or highest list item.
NULL	If NULL is specified, the global default is used. This global default is set initially to **packing.low**.
value	A scalar expression or list of scalars that provides a value that is to be unpacked.
target1, target2	One or more expressions separated by commas. Each expression is a path to a scalar or a compound data item, such as a struct, field, list, or variable.

Example 9-3 shows the usage of the **unpack()** method.

Example 9-3 Usage of unpack() Method

```
Example shows the usage of the unpack() method
to unpack a list of bits or bytes into a destination format.
There are two options most commonly used with
the unpack() method:
- packing.high: Starts filling the fields into higher
order positions.
- packing.low: Starts filling the fields into lower
order positions.
```

Example 9-3 Usage of unpack() Method (Continued)

```
<'
struct instruction { //Define target struct
    %opcode      : uint (bits : 3); //Target struct field
    %operand     : uint (bits : 5); //Target struct field
    %address     : uint (bits : 8); //Target struct field
};

extend sys {
    post_generate() is also {
        var inst : instruction; //Variable of struct instruction
        var packed_data: list of bit; //Source list of bit
        packed_data = {1;1;1;1;0;0;0;0;1;0;0;1;1;0;0;1};
        //This is equivalent to 1001_1001_0000_1111 with
        //1111 being the least significant nibble

        unpack(packing.high, packed_data, inst);
        //This function unpacks as follows
        //opcode == 100
        //operand == 1_1001
        //opcode == 0000_1111
        //packed_data is the source list of bytes
        //inst is the target struct instance
    };
};
'>
```

Figure 9-2 shows the order in which the source list *packed_data* is unpacked onto the fields of the variable *inst*.

Figure 9-2 Order of Unpacking

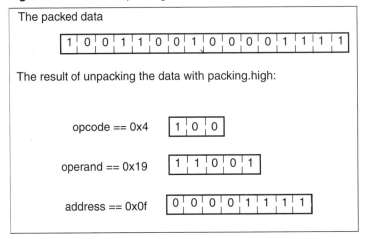

9.2 Data Checking

There are two aspects to checking in *e*, data checks and temporal checks. This section covers facilities for data checking.

9.2.1 Comparing Two Structs for Equality

Using the **pack()** and **unpack()** methods, structs can be converted to a list of bits or bytes and vice versa. Therefore, it is recommended that all verification be done in *e* code by means of abstract structs. The **pack()** method should be used to convert the structs to a list of bits at the time of driving. Similarly, the **unpack()** method should be used to convert the list of bits received from the DUT into an abstract struct. Finally, the abstract structs should be compared for equality instead of bits. Figure 9-3 shows the methodology for the comparison of these structs.[2] This methodology is also recommended for simplifying debugging of miscomparisons.

---

2. The methodology for struct comparison is just one example. There are many other ways of comparing for equality. For example, in a CPU verification system, a reference model is used and the output of that model is compared with the DUT for equality. Although the methodology may change, usage of methods described in this section still applies.

Figure 9-3 Comparing Input and Output Structs

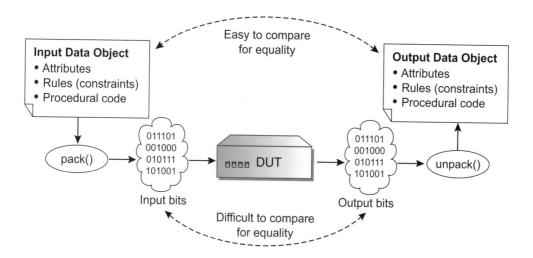

In *e*, two structs can be compared field by field to check for mismatches. As shown in Figure 9-3, this is very useful when the input struct and the output struct are compared. There are two routines that compare structs for field-by-field equality, **deep_compare()** and **deep_compare_physical()**.

9.2.1.1 deep_compare_physical()

This routine recursively compares each physical field of a struct instance with the corresponding field of another struct instance. Virtual fields are ignored. If there are multiple levels of instantiation hierarchy, then **deep_compare_physical()** compares at these levels. The syntax for **deep_compare_physical()** is as follows:

```
deep_compare_physical(struct-inst1: exp, struct-inst2: exp, max-diffs:
int): list of string;
```

Example 9-4 shows the usage of the **deep_compare_physical**() routine.

Example 9-4 deep_compare_physical() Routine

```
Example shows the usage of the deep_compare_physical() routine.
<'
struct packet { //Define a struct
    %header: header; //Physical field, header is a struct
    %data[10] :list of byte; //Physical field
    protocol: [ATM, ETH, IEEE]; //Virtual field
};

struct header {
    %code: uint; //Physical field
                 //deep_compare_physical() will go into
                 //header struct and do a comparison of code field
};

extend sys {
    pmi[2]: list of packet; //Generate two packet instances
    post_generate() is also {
        var diff: list of string;
        //Call to deep_compare_physical()
        diff = deep_compare_physical(pmi[0], pmi[1], 10);
             //Compare the physical fields of pmi[0] and pmi[1] for
            //equality. This is a recursive comparison. It will
          //compare fields, code and data. However, it
          //will ignore the protocol
          //field. Output is a list of string mismatches.
            //The number 10 is the max mismatches to be reported.
        if (diff.size() != 0) { //If there are mismatches
            out(diff); //print out the mismatches.
        }; //Otherwise, if there are no mismatches, a null
            //string will be printed, meaning the packets matched
          //successfully.
    };
};
'>
```

9.2.1.2 deep_compare()

The **deep_compare**() routine recursively compares both virtual and physical fields of a struct instance with the corresponding fields of another struct instance. If there are multiple levels of instantiation hierarchy, then **deep_compare()** compares at these levels. The syntax for **deep_compare()** is as follows:

```
deep_compare(struct-inst1: exp, struct-inst2: exp, max-diffs: int):
list of string;
```

Example 9-5 adds to the code in Example 9-4 to show the usage of the **deep_compare()** routine.

Example 9-5 deep_compare() Routine

```
<'
extend sys {
post_generate() is also {
        var diff1: list of string;
        //Call to deep_compare()
        diff1 = deep_compare(pmi[0], pmi[1], 10);
            //Compare both physical and virtual fields of
            //pmi[0] and pmi[1] for
            //equality. This is a recursive comparison. It will
            //compare the fields, code, data and protocol.
            //Output is a list of string mismatches.
            //The number 10 is the max mismatches to be reported.
        if (diff1.size() != 0) { //If there are mismatches
           out(diff1); //print out the mismatches.
        }; //Otherwise, if there are no mismatches, a null
           //string will be printed, meaning the packets matched
           //successfully.
    };
};
'>
```

9.2.2 Check that Action

The **check that** action performs a data comparison and, depending on the results, prints a message. The keyword **that** is optional. The **check that** action is functionally equivalent to the if-then-else construct but is recommended for checks because it provides better clarity in documenting checking code. The syntax for the **check that** construct is as follows:

```
check [that] bool-exp [else dut_error(message: exp, ...)];
```

Example 9-6 shows the usage of the **check that** action.

Example 9-6 Check that Action

```
Example shows the usage of the check that action.
This is primarily used for data checking.
<'
extend sys {
check_hard_error() is {
    check that 'top.hard_error' == 1 else //boolean check
        dut_error("Error-5 -- Hard error not asserted");
};

//If the above boolean check does not match, an error message
// as shown below is displayed.
//*** Dut error at time 0
// Checked at line 4 in check2.e
// In sys-@0.check_hard_error():
//
// Error-5 -- Hard error not asserted
//
//Will stop execution immediately (check effect is ERROR)
//
// *** Error: A Dut error has occurred

//If the else dut_error() clause is omitted, a default
//message is displayed
'>
```

9.2.3 dut_error() Method

This method is used to specify a DUT error message string. This action is usually associated with an **if** action, a **check that** action, or an **expect** struct member. If the boolean expression in the associated action or struct member evaluates to TRUE, then the error message string is displayed. The syntax for the **dut_error**() method is as follows:

```
dut_error(message: exp, ...);
```

Example 9-7 shows the usage of the **dut_error**() method.

Example 9-7 dut_error() Method

```
Example shows the usage of the dut_error() method.
<'
extend sys {
    m() is {
        if 'data_out' != 'data_in' then
            {dut_error("DATA MISMATCH: Expected ", 'data_in')};
    };
};


//If there is a mismatch, the following message
//will be printed on the screen
//*** Dut error at time 0
// Checked at line 4 in /tests/check6.e
// In sys-@0.m():
//
//DATA MISMATCH: Expected 1
//
//Will stop execution immediately (check effect is ERROR)
//
// *** Error: A Dut error has occurred
//
'>
```

9.2.4 Check Failure Effects

Often, you may want your error/debug messages to behave differently in module testing, system testing and regression. *e* allows you to change the behavior of error/debug messages without touching the original *e* code. This can be done either from an extension of the existing *e* code or from Specman Elite. Check failure effects can be set by use of the **set_check**() routine from *e* code or the **set check** command from the Specman Elite prompt. The syntax for the **set_check**() routine is as follows:

```
set_check(static-match: string, check-effect: keyword);
```

Table 9-3 describes the components of a **set_check**() routine.

Table 9-3 Components of set_check() Routine

static-match	A regular expression enclosed in double quotes. Only checks whose message string matches this regular expression are modified. Use … to match white space or non-white space characters.

check-effect is one of the following:

ERROR	Specman Elite issues an error message, increases **num_of_dut_errors**, breaks the run immediately and returns to the simulator prompt.
ERROR_BREAK_RUN	Specman Elite issues an error message, increases **num_of_dut_errors**, breaks the run at the next cycle boundary and updates the simulation GUI with the latest values.
ERROR_AUTOMATIC	Specman Elite issues an error message, increases **num_of_dut_errors**, breaks the run at the next cycle boundary, and performs the end of test checking and finalization of test data that is normally performed when **stop_run()** is called.
ERROR_CONTINUE	Specman Elite issues an error message, increases **num_of_dut_errors**, and continues execution.
WARNING	Specman Elite issues a warning, increases **num_of_dut_warnings** and continues execution.
IGNORE	Specman Elite issues no messages, does not increase **num_of_-dut_errors** or **num_of_dut_warnings,** and continues execution.

Example 9-8 shows the usage of the **set_check()** routine.

Example 9-8 Usage of set_check() Routine

```
Example shows how to extend the setup() method
of sys to set up check effects. If the e code
shown below is loaded, all dut_error() messages
in that simulation will be shown as WARNING.
<'
extend sys {
    setup() is also {
        set_check("...", WARNING);
    };
};
'>
```

Example 9-8 Usage of set_check() Routine (Continued)

```
//The check effect can also be set from the
//the Specman Elite prompt.
//Specman> set check "..." WARNING
```

9.3 Temporal Checking

Checking can be performed to test the success or failure of any simple or complex temporal expression. This section discusses temporal checking by means of *e* struct members **expect** and **on**. Temporal checking is often referred to as assertion checking.

9.3.1 Expect Struct Member

The **expect** struct member defines temporal rules. If the temporal expression fails at its sampling event, the temporal rule is violated and an error is reported. If there is no **dut_error()** clause, the rule name is displayed. Once a rule has been defined, it can be modified using the **is only** syntax. The syntax for the **expect** struct member is shown below:

```
expect [rule-name is [only]] temporal-expression
    [else dut_error(string-exp)];
```

Table 9-4 shows the components of an **expect** struct member.

Table 9-4 Components of an expect Struct Member

rule-name	An optional name that uniquely distinguishes the rule from other rules or events within the struct. You can use this name to override the temporal rule later.
temporal-expression	A temporal expression that is always expected to succeed. Typically involves a temporal yield (=>) operation.
string-exp	A string or a method that returns a string. If the temporal expression fails, the string is displayed, or the method is executed and its result is displayed.

Example 9-9 shows the usage of an **expect** struct member.

Example 9-9 Expect Struct Member

```
This example defines an expect rule.
This rule requires that the length of the bus cycle be no longer than
1000 cycles. This is the number of clocks between the transmit_start
and transmit_end event.
<'
struct bus_e {
    event bus_clk is change('top.b_clk') @sim; //Event
    event transmit_start is rise('top.trans') @bus_clk; //Event
    event transmit_end is rise('top.transmit_done') @bus_clk; //Event
    event bus_cycle_length; //Declare event
    expect bus_cycle_length is //Set up a rule for that event
        //If transmit_start occurs, transmit_end must occur
        //within 1000 cycles of @bus_clk
        @transmit_start => {[0..999];@transmit_end} @bus_clk
        else dut_error("Bus cycle did not end in 1000 cycles");
};
'>

//If the bus cycle is longer than 1000 cycles, the following message
//will be issued.
//-------------------------------------------------------
// *** Dut error at time 1000
// Checked at line 7 in @expect_msg
// In bus_e-@0:
//
//bus_cycle_length: Bus cycle did not end in 1000 cycles
//-------------------------------------------------------
//Will stop execution immediately (check effect is ERROR)
//
// *** Error: A Dut error has occurred
```

9.3.2 On Struct Member

The **on** struct member executes a block of actions immediately whenever a specified trigger event is emitted. An **on** struct member is similar to a regular method except that the action block for an **on** struct member is invoked immediately upon the emission of the trigger event. An **on** action block is executed before TCMs waiting for the same event.

The **on** action block is invoked every time the trigger event is emitted. The actions are executed in the order in which they appear in the action block. You can extend an **on** struct member by repeating its declaration, with a different action block. This has the same effect as using **is also** to extend a method. The **on** action block cannot contain any TCMs.

The syntax of the **on** struct member is as follows:

```
on event-type {action; ...}
```

Example 9-10 shows the usage of an **on** struct member.

Example 9-10 On Struct Member

```
Example shows the usage of the on struct member.
<'
struct cnt_e {
  event ready;
  event start_count;

  //On struct member
  on ready {sys.req = 0}; //When ready is emitted, set sys.req = 0.
  //On struct member
  on start_count { //When start_count is emitted, perform actions.
    sys.count = 0;
    sys.counting = 1;
    outf("Starting to count - sys.count = %d\n", sys.count);
  };


  trigger()@sys.any is { //TCM that triggers the events for on struct
                         //member.
    wait cycle; //Wait for next occurrence of sys.any.
    emit ready; //Trigger ready.
    wait cycle; //Wait for next occurrence of sys.any.
    emit start_count; //Trigger start_count.
    wait cycle; //Wait for next occurrence of sys.any.
    stop_run();
  };
  run() is also {
    start trigger();
  };
};
'>
```

9.4 Summary

- The *e* language syntax provides a way of tagging physical fields by prefixation of these fields with a **%** symbol. Physical fields are significant only for the **pack()** and the **unpack()** methods. Otherwise, their behavior is identical to that of virtual fields.

- Packing (**pack**() method) is commonly used to prepare high-level *e* data into a form that can be applied to a DUT. Packing performs concatenation of items, including items in a list or fields in a struct, in the order specified by the pack options parameter and it returns a list of bits.

- The **unpack**() method does exactly the opposite of packing. Unpacking is commonly used to convert a raw bit or byte stream into high-level data by storing the bits of the value expression into the target expressions. Unpacking operates on scalar or compound (struct, list) data items.

- There are two routines that compare structs for field-by-field equality, **deep_compare**() and **deep_compare_physical**().

- The **check that** action performs a data comparison and, depending on the results, prints a message. The **that** keyword is optional.

- The **dut_error**() method is used to specify a DUT error message string. This action is usually associated with an **if** action, a **check that** action, or an **expect** struct member. If the boolean expression in the associated action or struct member evaluates to TRUE, then the error message string is displayed.

- Check failure effects can be set by use of the **set_check**() routine from *e* code or **set check** command from the Specman Elite prompt.

- The **expect** struct member defines temporal rules. If the temporal expression fails at its sampling event, the temporal rule is violated and an error is reported.

- The **on** struct member executes a block of actions immediately whenever a specified trigger event is emitted. The **on** action block is invoked every time that trigger event is emitted. The actions are executed in the order in which they appear in the action block.

9.5 Exercises

1. Define a struct *packet*. Define the following elements in this struct.

    a. *addr* (2 bits) - physical field

    b. *len* (6 bits) - physical field

    c. *payload* (list of byte) - physical field

    d. *len* field equals the length of the payload (constraint)

    e. *trailer* (byte) - physical field

    f. Constrain each byte of *payload* to be equal to the index of that byte.

2. Extend **sys**. Create an instance *packet1* of type *packet* in **sys**. Create an instance *packet2* of type *packet* in **sys**. Mark the *packet2* instance as do-not-generate (!).

   a. Extend the **run()** method of **sys**.

   b. Declare a local variable *packed_data* as list of byte.

   c. Print *packet1*.

   d. Pack the *packet1* instance to *packed_data* using **packing.high**.

   e. Print the value of *packed_data*.

   f. Unpack the *packed_data* to the *packet2* instance.

   g. Print *packet2*.

3. Extend **sys**.

   a. Extend the **run()** method of **sys**.

   b. Create a local variable *diff* that is of type list of string.

   c. Compare the physical fields of *packet1* and *packet2* and store the result in *diff*. (Hint: Use **deep_compare_physical()**.)

   d. Check that *diff* is empty, otherwise print *diff* with **dut_error()**. (Hint: Use **check that** syntax and **is_empty()** method.)

   e. Compare all fields of *packet1* and *packet2* and store the result in *diff*. (Hint: Use **deep_compare()**.)

   f. Check that *diff* is empty, otherwise print *diff* with **dut_error()**. (Hint: Use **check that** syntax and **is_empty()** method.)

4. Extend **sys**. Extend the **setup()** method to set the check effect of all messages to WARNING.

5. Write the **expect** struct members for the following scenarios. $@p\_clk$ is the sampling event.

   a. Declare events *p_clk*, *req*, *ack*, *p_start*, and *p_end*.

   b. If the $@req$ event occurs, $@ack$ should occur within 5 to 10 cycles after $@req$.

   c. If the $@p\_start$ event occurs, $@ack$ should occur within 5 to 10 cycles after $@p\_start$ followed by $@p\_end$ within 25 to 100 cycles of $@ack$.

6. Extend **sys**. Declare a *count* field in **sys**. Write the following **on** struct members within **sys**.

   a. On the occurrence of the *req* event, print the message "Request event occurred".

   b. On the occurrence of the *p_start* event, print the message "Packet started". Set the field *count* to the value 0.

CHAPTER 10

Coverage

Even in the most sophisticated verification environment, the question is, how much verification is enough? The answer to this question can only be provided by coverage. Coverage also provides answers about where most energies of the verification engineer should be focused. Coverage constructs in *e* provide a means of integrating coverage as a part of basic verification methodology. This approach maximizes stimulus generation efficiency, maximizes the design coverage, and minimizes the number of repetitive tests. This chapter discusses the coverage constructs provided by *e*.

Chapter Objectives

- Describe functional coverage.
- Define coverage groups and coverage group options.
- Describe coverage items and coverage item options.
- Explain transition coverage items.
- Describe cross coverage items.
- Understand latency coverage.
- Explain how to set up coverage.

10.1 Functional Coverage

Functional coverage perceives the design from a user's or a system point of view. It monitors whether all important stimulus scenarios, error cases, corner cases, and protocols are covered in the verification system. Functional coverage also measures if all important combinations of input stimulus have been exercised at different states of the DUT.

Functional coverage elevates the discussion to specific transactions or bursts without overwhelming the verification engineer with bit vectors and signal names. Therefore, the bulk of low-level details are hidden from the report reviewer. This level of abstraction enables natural translation from measured coverage to generation constraints.

Thus, functional coverage is very important because:

- It allows the verification engineer to focus on key areas of the design that need the most attention.
- It tells the verification manager how much verification is enough.
- It improves the efficiency of stimulus generation.
- It improves the quality of stimulus.
- It avoids repetitive generation of stimulus for the same set of combinations.

There are other types of coverage such as code coverage, toggle coverage, etc. measured by other tools. These coverage options are needed in addition to functional coverage. The following sections focus on the functional coverage constructs in *e*.

10.2 Coverage Groups

A coverage group is a struct member defined by the **cover** keyword. A coverage group contains a description of data items for which data are collected over time. A coverage group has an associated **event** that tells when to sample and collect data. The **event** must be declared in the same struct as the coverage group. The syntax for a coverage group definition is as follows:

```
cover event-type [using coverage-group-option, ...] is {coverage-item-
definition; ...};
```

Table 10-1 defines components of a coverage group definition.

Table 10-1 Components of a Coverage Group Definition

event-type	The name of the group. This must be the name of an event type defined in the same struct. The event must not have been defined in a subtype.
	The event is the sampling event for the coverage group. Coverage data for the group are collected every time the event is emitted.
	The full name of the coverage group is *struct-exp.event-type*. The full name must be specified for the **show cover** command and other coverage commands and methods.
coverage-group-option	Each coverage group can have its own set of options. The options can appear in any order after the **using** keyword. The coverage group options listed in Table 10-2 can be specified with the **using** keyword.

Table 10-1 Components of a Coverage Group Definition (Continued)

coverage-item-definition	The definition of a coverage item.
is also	Coverage groups can be extended like other struct members.
is empty	The **empty** keyword can be used to define an empty coverage group that will be extended later, using a **cover is also** struct member with the same name.

Table 10-2 shows options while defining coverage groups.

Table 10-2 Coverage Group Options

Option	Description
no_collect	This coverage group is not displayed in coverage reports and is not saved in the coverage files. This option enables tracing of coverage information and enables event viewing with **echo event**, without saving the coverage information.
count_only	This option reduces memory consumption because the data collected for this coverage group are reduced. You cannot do interactive, post-processing cross coverage of items in **count_only** groups. The coverage configuration option **count_only** sets this option for all coverage groups.
text=*string*	A text description for this coverage group. This can only be a quoted string, not a variable or expression. The text is shown at the beginning of the information for the group in the coverage report (displayed with the **show cover** command).
when=*bool-exp*	The coverage group is sampled only when *bool-exp* is TRUE. The *bool-exp* is evaluated in the context of the parent struct.
global	A global coverage group is a group whose sampling event is expected to occur only once. If the sampling event occurs more than once, Specman Elite issues a DUT error. If items from a global group are used in interactive cross coverage, no timing relationships exist between the items.

Table 10-2 Coverage Group Options (Continued)

Option	Description
radix=DEC|HEX| BIN	Buckets for items of type **int** or **uint** are given the item value ranges as names. This option specifies in which radix the bucket names are displayed. A bucket is a group of different values of a coverage item collected for coverage.
	The global **print** radix option does not affect the bucket name radix.
	Legal values are DEC (decimal), HEX (hexadecimal), and BIN (binary). The value must be in uppercase letters.
	If the **radix** is not used, **int** or **uint** bucket names are displayed in decimal.
weight=*uint*	This option specifies the grading weight of the current group relative to other groups. It is a nonnegative integer with a default of 1.

Example 10-1 shows sample coverage definitions that use coverage group options.

Example 10-1 Coverage Group Definition

```
Example shows coverage group definition.
< '
type cpu_opcode: [ADD, SUB, OR, AND, JMP, LABEL];
struct inst {
    reset_done: bool; //Set by some struct in sys. Not shown.
    opcode: cpu_opcode; //Define enumerated type field.
    good_opcode: bool; //Define boolean.
    event info; //Event that defines coverage group name.
                //Coverage samples are collected when the
                //data_change event is emitted.
```

Example 10-1 Coverage Group Definition (Continued)

```
    cover info using //Define coverage group
              count_only, //count_only option.
              radix = HEX, //Display in hex format.
              weight = 10, //Relative weight compared to
                            //other groups.
              when = (reset_done == TRUE) //Collect coverage only
                                          //when this bool
                                          //expression is true.
    is {
        item opcode; //Item opcode is to be covered.
    };

    run() is also { //Emit the coverage events in the run() method.
         emit info; //Emit the event that causes coverage samples to
                    //be taken.
    };}; //end of struct inst


type cpu_state: [START, FETCH1, FETCH2, EXEC]; //Enumerated type.
struct cpu {
    init_complete: bool; //Set by some struct in sys. Not shown.
    event state_change; //Define event for coverage group.
    cover state_change using //State machine coverage group.
           text = "Main state-machine", //Name of coverage group.
           weight = 5, //Relative weight
           when = (init_complete == TRUE) //Collect sample only if
                                          //bool exp is true.
    is {
    //Coverage of HDL state machine state variable.
    item st: cpu_state = '~/top/cpu/main_cur_state';
    };
    run() is also { //Emit the coverage events in the run() method.
        emit state_change; //Emit the event that causes coverage
    };                     //samples to be taken.

};
'> End of e code
```

10.2.1 Extending Coverage Groups

A coverage group can also be extended like any other struct member by means of the **is also** keyword. Example 10-2 shows the extension of the coverage group defined in Example 10-1.

Example 10-2 Extending Coverage Groups

```
Example shows how to extend coverage groups
<'
extend inst {
    cover info is also { //Extend the coverage group.
        item good_opcode; //Item good_opcode is also to be covered
                          //in addition to item opcode.
    };
}; //end of extension of struct inst
'>
```

There are three types of items in a coverage group: basic coverage items, transition items, and cross items. Figure 10-1 shows the three types of items in a coverage group. These items are discussed in detail in the following sections.

Figure 10-1 Coverage Group Items

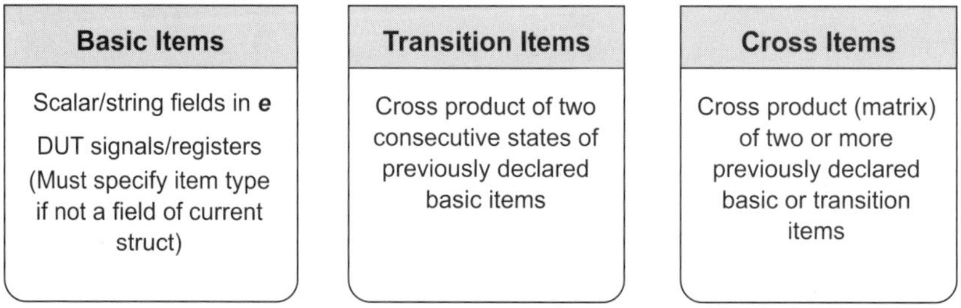

Basic Items	Transition Items	Cross Items
Scalar/string fields in **e** DUT signals/registers (Must specify item type if not a field of current struct)	Cross product of two consecutive states of previously declared basic items	Cross product (matrix) of two or more previously declared basic or transition items

10.3 Basic Coverage Item

A basic coverage item can be one of the following three elements.

- A field in the local *e* struct
- A field in another *e* struct in the hierarchy
- An HDL signal or register

A basic coverage item can be specified with options that control how coverage data are collected and reported for the item. The item can be an existing field name or a new name. If you use a new name for a coverage item, you must specify the item's type and the expression that defines it. The syntax for a basic coverage group item is as follows:

```
item item-name[:type=exp] [using coverage-item-option, ...];
```

Table 10-3 describes the important coverage item options. These options can be specified with the definition of each basic coverage item.

Table 10-3 Basic Coverage Item Options

Option	Description
per_instance	Coverage data are collected and graded for all the other items in a separate listing for each bucket of this item.
text=_string_	A text description for this coverage item. This can only be a quoted string, not a variable or expression.
when=_bool-exp_	The item is sampled only when _bool-exp_ is TRUE. The _bool-exp_ is evaluated in the context of the parent struct.
at_least=_num_	The minimum number of samples for each bucket of the item. Anything less than _num_ is considered a hole.
ranges = {**range**(_parameters_);...}	Create buckets for this item's values or ranges of values.
ignore=_item-bool-exp_	Defines values that are to be completely ignored. They do not appear in the statistics at all. The expression is a boolean expression that can contain a coverage item name and constants.
illegal=_item-bool-exp_	Defines values that are illegal. An illegal value causes a DUT error.
radix=DEC|HEX|BIN	Specifies the radix used to display coverage item values.
weight=_uint_	Specifies the weight of the current item relative to other items in the same coverage group. It is a non-negative integer with a default of 1.
name	Assigns an alternative name for a cross or transition item.

Example 10-3 illustrates basic coverage items.

Example 10-3 Basic Coverage Items

```
Example shows three types of basic coverage items:
- Fields in a local struct
- Fields in another struct
- HDL signals
An item must have a name if it is assigned to a field
in another struct or an HDL signal. For fields in a local
struct, the field name can be the item name.
See the item options with each item.

<'
type cpu_opcode: [ADD, SUB, OR, AND, JMP, LABEL];
type cpu_reg: [reg0, reg1, reg2, reg3];
struct inst {
    opcode: cpu_opcode;
    op1: cpu_reg;
    op2: byte;
    event inst_driven;
    cover inst_driven is { //Three types of items
        //Field op1 in local struct, see item options
        item op1 using weight = 5, radix = DEC;
        //Field op2 in local struct, see item options
        item op2 using ignore = 0, radix = HEX, at_least = 10;
        //Name op2_big equal to boolean expression, see item options
        item op2_big: bool = (op2 >= 64) using weight = 10;
        //Name op3 equal to field in another struct
        item op3: bool = sys.header.op3;
        //Name hdl_sig equal a HDL variable value
        item hdl_sig: int = '~/top/sig_1';
    };
};
'>
```

10.3.1 Basic Coverage Item Ranges

When one is covering **int/uint** types, it is not practical or necessary to cover the entire range of values for a 32-bit integer. Therefore, it is necessary to create a bucket for each boundary value or a range of values. Buckets can be created by means of the **ranges** option. Each bucket counts the number of samples within its range. The syntax for the **ranges** option is as follows:

```
ranges = {range(parameters); range(parameters); ...}
range(range: range, name: string, every-count: int, at_least-num: int);
```

The arguments to **range()** are shown in Table 10-4 below.

Table 10-4 Arguments to range() for Basic Coverage Item

range	The range for the bucket. It must be a literal range such as "[1..5]", of the proper type. Even a single value must be specified in brackets (for example "[7]"). If you specify ranges that overlap, values in the overlapping region go into the first of the overlapping buckets. The specified range for a bucket is the bucket name. That is, the buckets above are named "[1..5]" and "[7]".
name	A name for the bucket. If you use the *name* parameter, you cannot use an *every-count* value. You must enter UNDEF for the *every-count* parameter.
every-count	The size of the buckets to create within the range. These buckets get automatically created by the *every-count* parameter. If you use the *every-count* parameter, you cannot use a *name*. You must enter an empty string ("") as a placeholder for the *name* parameter.
at-least-num	A number that specifies the minimum number of samples required for a bucket. If the item occurs fewer times than this, a hole is marked. This parameter overrides the global **at_least** option and the per-item **at_least** option. The value of *at-least-num* can be set to zero, meaning "do not show holes for this range."

Example 10-4 illustrates the **ranges** option.

Example 10-4 Ranges Option for Basic Coverage Item

```
Example of basic coverage group item using the ranges option
<'
struct pcc {
    pc_on_page_boundary: uint (bits: 15); //Local field
    pc: uint (bits: 15); //Local field
    stack_change: byte; //Local field
```

Example 10-4 Ranges Option for Basic Coverage Item (Continued)

```
    event pclock; //Coverage group event
    cover pclock is { //Define coverage group
        item pc_on_page_boundary using //Cover the local field
            ranges = { //Ranges option
                range([0], "0"); //Only one value (0) in bucket
                range([4k], "4k"); //Only one value (4k) in bucket
                range([8k], "8k"); //Only one value (8k) in bucket
                range([12k], "12k"); //Only one value (12k) in bucket
                range([16k], "16k"); //Only one value (16k) in bucket
                range([20k], "20k"); //Only one value (20k) in bucket
                range([24k], "24k"); //Only one value (24k) in bucket
                range([28k], "28k"); //Only one value (28k) in bucket
                range([0..32k-1], "non-boundary"); //Default buckets
            };

        item pc using radix = HEX, //Cover the local field
        ranges = { //Ranges option
                //range is 0..4k-1, name = page_0
                //Cannot use every_count(UNDEF) due to name for range
                //at_least = 4 samples needed in this bucket
                range([0..4k-1], "page_0", UNDEF, 4);
                //range is 4k..32k-1, no name
                //every_count = 8k, creates buckets of 8k size
                //at_least = 2 samples needed in this bucket
                range([4k..32k-1], "", 8k, 2);
            };
        item stack_change using //Cover the local field
            //Range is 0..32 divided into buckets of size 1
            //i.e. 33 buckets
            ranges = { range( [0..32], "", 1); };
    };
    run() is also {
        emit pclock; //Collect coverage
    };
};
'>
```

10.4 Transition Coverage Items

Transition coverage items record the change in the item's value between two consecutive samplings of the item. Transition items are very useful for covering DUT state machines. The syntax for the transition item is as follows:

```
transition item-name [using coverage-item-option, ...];
```

The coverage options for transition coverage items are very similar to the options described in Table 10-3 for basic coverage items. However, the **ranges** option is not available with transition coverage items. The transition between two samples is specified with the key word **prev_** that is prefixed before the name of the transition coverage item. Example 10-5 shows the usage of transition coverage items for covering a DUT state machine.

Example 10-5 Transition Coverage Items

```
Example that shows how to cover a DUT state machine
using transition items.
<'
type cpu_state: [START, FETCH1, FETCH2, EXEC];
struct cpu {
    //Define a coverage event
    event clk_rise is rise('~/top/clk') @sim;
    //Define a coverage struct member

    cover clk_rise is {
         //Define item that tracks the HDL state vector
        item state: cpu_state = '~/top/state';
        //Define transition items using the option
        //illegal. This option prints an error message
        //when the boolean expression is true. Inside the
        //boolean expression is a list of all legal state
        //transitions with a not() around it i.e all illegal
        //state transitions. Thus, if any illegal transition
        //is taken, an error message is displayed. This technique
        //is preferred compared to specifying all illegal transitions.
        transition state using illegal =
            //A not of all legal state transitions
            not ((prev_st == START and st == FETCH1)//State transition
            or (prev_st == FETCH1 and st == FETCH2)//State transition
            or (prev_st == FETCH1 and st == EXEC)//State transition
            or (prev_st == FETCH2 and st == EXEC)//State transition
            or (prev_st == EXEC and st == START)); //State transition
    };
};
'>
```

10.5 Cross Coverage Items

Cross coverage items provide a cross product (matrix) of two or more previously declared basic items. It is possible to create a cross of any number of basic or transition items declared in the same coverage group. A cross can be created either from *e* code or from the coverage user interface of Specman Elite. The syntax for cross coverage items is as follows:

```
cross item-name-1, item-name-2, ... [using coverage-item-option, ...];
```

The coverage options for transition coverage items are very similar to the options described in Table 10-3 for basic coverage items. However, the **ranges** option is not available with cross coverage items. Example 10-6 shows the usage of transition coverage items for covering a DUT state machine.

Example 10-6 Cross Coverage Items

```
Example of cross coverage items.
A cross can be done from
- basic coverage items
- transition items
<'
type cpu_opcode: [ADD, SUB];
type cpu_reg: [reg0, reg1, reg2];
struct inst {
    opcode: cpu_opcode;
    op1: cpu_reg;
    event inst_driven;

    cover inst_driven is {
        item opcode; //Basic coverage item
        item op1; //Basic coverage item
        cross opcode, op1; //Cross between op1 and opcode
                          //Covers eight possible combinations of
                          //cpu_opcode and cpu_reg

    };
    run() is also {
        emit inst_driven;
    };
};
'>
```

10.6 Latency Coverage

Latency coverage requires comparison of the time an activity ends with the time that activity started. There are no built-in constructs provided by *e* for latency coverage. However, it is easily possible to perform latency coverage using *e*.

To perform latency coverage, do the following steps:

- When the activity begins, log the time.
- When the activity ends, subtract the logged time from the current time.
- Set up coverage on the time difference value.

Example 10-7 shows how to record latency coverage.

Example 10-7 Latency Coverage

```
Example of latency coverage
<'
struct packet {
    !gen_time: time; //Snapshot of time value
    !latency: time; //Value of time difference
    event clk is rise ('~/top/clk') @sim; //Define posedge of clock
    event pkt_started is rise('~/top/packet_vld')@clk; //Packet start
    event pkt_ended is fall('~/top/packet_end')@clk; //End of packet

    //Take a snapshot of system time on start of packet
    //sys.time gets the value of current simulation time
    on pkt_started { gen_time = sys.time };

    //Compute the value of latency at end of packet
    on pkt_ended { latency = sys.time - gen_time};

    cover pkt_ended is { //Set up coverage when the packet finishes
        item latency using ranges = { //Range values for latency
            range([1..20], "short"); //Short latency
            range([21..40], "medium"); //Medium latency
            range([41..63], "long"); //Long latency
        };
    };
}; //end of struct packet
'>
```

10.7 Turning On Coverage

Use the cover configuration option to turn on coverage, as in Example 10-8 below. If cover groups are defined then coverage is automatically turned on. Note that in this example, although only one mode is illustrated, other modes for coverage are available. These modes for coverage are discussed in detail in the *e* language reference manuals.

Example 10-8 Turning on Coverage

```
Example shows how to turn on the coverage
for the various coverage groups defined in the
e code. The set_config() method is called to turn
on coverage. The set_config() method has many different
options. The simplest option is shown in this example.
This file should be compiled with other e files to turn
on coverage.

<'
extend sys {
    setup() is also { //Extend the setup() method of sys.
        set_config(cover, mode, on); //Turns on coverage
                                     //Other modes are available
    };
};
'>
```

10.8 Viewing Coverage Results

After the *e*-based simulation is completed, coverage output is produced. Coverage output from Specman Elite can be viewed graphically or in a text file. Coverage output can also be accumulated over multiple test runs to view cumulative coverage. The details of the graphical interface for viewing coverage results are not discussed in this book.

10.9 Summary

- Functional coverage perceives the design from a user's or a system point of view. It checks whether all typical scenarios, error cases, corner cases, and protocols are covered in the verification system. Functional coverage also measures if all important combinations of input stimulus have been exercised at different states of the DUT.

- A coverage group is a struct member defined by the **cover** keyword. A coverage group contains a description of data items for which data are collected over time. A coverage group has an associated **event** that tells when to sample data. The **event** must be declared in the same struct as the coverage group.

- Coverage groups can also be extended like any other struct members using the **is also** keyword.

- There may be three types of items in a coverage group: basic coverage items, transition items, and cross items.

- A basic coverage item can be one of the following three elements: a field in the local *e* struct, a field in another *e* struct in the hierarchy, or an HDL signal or register. A basic coverage item can be specified with options that control how coverage data are collected and reported for the item.

- Transition coverage items record the change in the item's value between two consecutive samplings of the item. Transition items are very useful for covering DUT state machines.

- Cross coverage items provide a cross product (matrix) of two or more previously declared basic items. It is possible to create a cross of any number of basic or transition items declared in the same coverage group. A cross can be created either from *e* code or from the coverage user interface of Specman Elite.

- Latency coverage requires comparison of the time an activity ends with the time that activity started. There are no built-in constructs provided by *e* for latency coverage. However, it is easily possible to perform latency coverage by use of *e*.

10.10 Exercises

1. Define a struct *packet*. Define the following elements in this struct.

  a. Define *addr* (2 bits)—physical field.

  b. Define *len* (6 bits)—physical field.

  c. Define *payload* (list of byte)—physical field.

  d. Define *new_packet*—event.

  e. Define coverage group *new_packet* that includes coverage on *addr* and *len*.

  f. Use a weight = 5, radix = HEX on *addr* and a weight = 10, radix = DEC on *len*.

2. Extend struct *packet*.

  a. Extend coverage group *new_packet* using **is also**.

  b. Add cross coverage on *addr* and *len*.

  c. Cross coverage name *addr_len*.

3. Define an enumerated type *fsm_state* with values [IDLE, EXEC, HOLD]. Extend struct *packet*.

  a. Extend coverage group *new_packet* using **is also**.

  b. Add an item *st* of type *fsm_state* that gets a value from HDL signal '~/top/state'.

  c. Add a transition item of *st*. Exclude illegal transitions. The legal state transitions are IDLE->EXEC, EXEC->HOLD, HOLD->EXEC, EXEC->IDLE. (Hint: Use a **not** of legal transitions.) Assign a name "*State_Machine*" to this item.

4. Extend struct *packet*.

    a. Add events *start* and *end* to the struct.

    b. Take a sample of **sys.time** when event *start* is emitted.

    c. Take a sample of **sys.time** when event *end* is emitted

    d. Extend coverage group *new_packet* using **is also**.

    e. Perform latency coverage on the simulation time difference between the *start* and *end*. Ranges for latency coverage are [0..50] = "short", [51..100] = "medium", [101..200] = "long". At least three samples are needed in each range for full coverage. (Hint: Use **ranges** keyword and *at_least* option.)

5. Extend **sys**. Instantiate the *packet* struct. Turn coverage ON in normal mode.

CHAPTER 11

Running the Simulation

So far in this book we have discussed different areas of *e* syntax. This chapter puts all the concepts together and discusses how to run a complete *e* simulation in conjunction with the HDL simulation. The concepts developed in this chapter are necessary to build and run complete *e*-based verification systems.

Chapter Objectives

- Describe verification components in detail.
- Define the evaluation hierarchy of **sys**.
- Describe the execution flow of Specman Elite.
- Understand the test phase methods of Specman Elite.
- Explain the synchronization between the HDL simulator and Specman Elite.

11.1 Verification Components

Multiple components make up a verification environment. Figure 11-1 shows the components in a typical verification environment. This section describes how various verification components are built and the *e* constructs that are needed to build these verification components.

Figure 11-1 Components of a Verification Environment

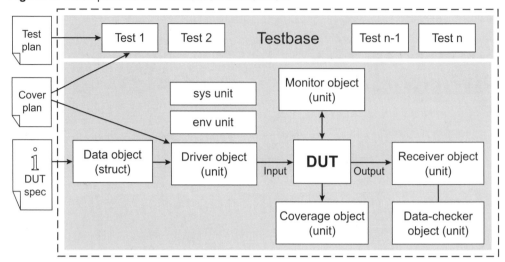

The information provided in this section is a guideline for creation of verification environments. Actual verification environments will vary slightly depending on the needs of the particular project.

11.1.1 Data Object (Stimulus Generation)

The generation aspect of verification contains *data objects*. These data objects are structs that represent the abstract form of input stimulus. Data objects can be built with the following *e* syntax.

- Data objects are declared as **struct**, not **unit**. Thus, data objects can be generated on-the-fly using the **gen** action.

- Constraints are defined using **keep** and **keep for each** constructs. Constraints ensure that legal stimulus is generated.

- Usually, there is one data object per input protocol. If the DUT handles multiple input protocols, each has a data object corresponding to the stimulus of that protocol.

11.1.2 Driver Object (Stimulus Driver)

The *driver* object performs the function of taking the stimulus data object one instance at a time and applying it to the DUT until all stimulus has been applied. Typically, the interface object is similar to an HDL bus functional model. In case of a networking environment, it is also known as a *port* object. Figure 11-2 shows the typical design of a driver object.

Figure 11-2 Driver Object Design

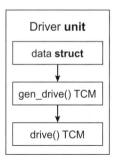

There can be more than one driver object in the design. Typically, there is one driver object per input interface. If there are multiple input protocols supported by the DUT, each has its own driver object. Here are some characteristics of driver objects:

- Driver objects are declared as **unit**. A **keep** constraint on the **hdl_path**() determines the HDL hierarchy instance to which that driver object is attached.

- Driver objects are typically written once per project.

- A *drive()* TCM with a sampling event defined as the clock of the input interface is responsible for driving one stimulus data object onto the input bus. The **pack**() method is used to convert an abstract data object into a list of bits or bytes.

- A *gen_drive()* TCM with a sampling event equal to the clock of the input protocol is responsible for generating many such stimulus data objects (using the **gen** action or a **list**) and applying them one by one to the DUT by calling the *drive()* TCM.

11.1.3 Receiver Object (Output Receiver)

The *receiver* object is responsible for collecting the data at the output of the DUT. The receiver object performs the function of taking the raw output data from the DUT based on the output protocol and converting it to an abstract data object for comparison. Typically, the receiver object is similar to an HDL bus functional model. Figure 11-3 shows the typical design of a receiver object.

Figure 11-3 Receiver Object Design

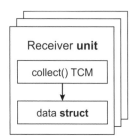

There can be more than one receiver object in the design. Typically, there is one receiver object per output interface. If there are multiple output protocols supported by the DUT, each has its own receiver object. In certain protocols, the driver and the receiver objects might be combined into one object. Here are some characteristics of receiver objects:

- Receiver objects are declared as **unit**. A **keep** constraint on the **hdl_path**() determines the HDL hierarchy instance to which that receiver object is attached.

- Receiver objects are typically written once per project.

- A *collect()* TCM with a sampling event equal to the clock of the output protocol is responsible for receiving one stimulus data object on the output bus. The **unpack()** method is used to convert a list of bits or bytes received on the output bus into an abstract data object.

- The output data object is then passed to the *data_checker* object for comparison against expected values.

11.1.4 Data Checker Object (Expected Value Comparison)

The *data_checker* object stores expected values for each input data object injected into the DUT. The expected values are computed by applicaion of the transformation algorithm to the input data objects. This transformation algorithm is identical to the one applied to the input stimulus in the DUT. When the receiver object receives one data object, it passes it to the *data_checker* for comparison against the expected values. Figure 11-4 shows the typical design of a *data_checker* object.

Figure 11-4 Data Checker Object Design

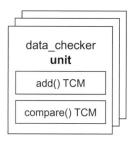

Data_checker objects may be instantiated in the receiver object. In such cases, there is one *data_checker* object per receiver object. Here are some characteristics of *data_checker* objects:

- *Data_checker* objects are declared as **unit**. A **keep** constraint on the **hdl_path**() determines the HDL hierarchy instance to which that *data_checker* object is attached.

- *Data_checker* objects are typically written once per project.

- An *add()* TCM with a sampling event equal to the clock of the output protocol is responsible for adding one stimulus data object received on the output bus for comparison against its expected value.

- A *compare()* TCM with a sampling event equal to the clock of the output protocol is responsible for comparing one stimulus data object received on the output bus against its expected value. The predefined **list** methods and **deep_compare_physical**() method are called to perform this comparison.

Scoreboarding is a common technique to perform such comparisons. Scoreboarding is a general verification technique. It is not specific only to *e*-based verification. Figure 11-5 shows a typical scoreboarding methodology that can be developed using *e*. In Figure 11-5, the data item is added to the scoreboard as soon as it is applied to the input of the DUT. When the output is received, the output is compared to the data item on the scoreboard.

Figure 11-5 Scoreboarding Methodology

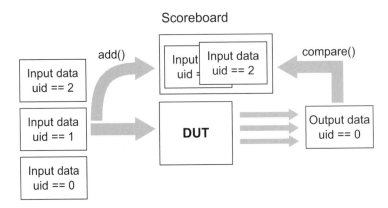

11.1.5 Monitor Object (Protocol Checker)

A *monitor* object checks the input and output timing protocol. Here are some characteristics of a monitor object:

- Monitor objects are declared as **unit**. A **keep** constraint on the **hdl_path()** determines the HDL hierarchy instance to which that monitor object is attached.
- Monitor objects are typically written once per project.
- Monitor objects are written with **expect** struct members.
- A **dut_error()** method is called if the **expect** fails.
- A monitor object may be built with multiple lower level objects.

11.1.6 Coverage Object (DUT and *e* Coverage)

A *coverage* object sets up coverage tracking on key items in the *e* code and in the DUT. Some coverage objects can also be embedded in the extension to a driver or receiver object. Here are some characteristics of a coverage object:

- Coverage objects are declared as **unit**. A **keep** constraint on the **hdl_path()** determines the HDL hierarchy instance to which that coverage object is attached.
- Coverage objects are typically written once per project.
- Coverage objects are written with **cover** struct members.

11.1.7 Tests

Tests are derived from the test plan. In *e*, tests are simply extensions of existing struct and unit definitions that impose additional constraints on these objects. Test files are very small and are easy to write and to maintain.

11.1.8 Env Object (Environment)

Often, a verification environment that is designed for a module will be needed at the chip level and at the system level. All verification components discussed in this section should be automatically transported to a different level of verification. Therefore, instead of directly instantiating these components in **sys**, create an intermediate level of hierarchy called *env* under **sys**. All verification components are instantiated under *env*. Thus, one can transport *env* to any other level of hierarchy without having to transport each individual verification component. Figure 11-6 shows the hierarchy with the *env* object.

Figure 11-6 Environment Object Hierarchy

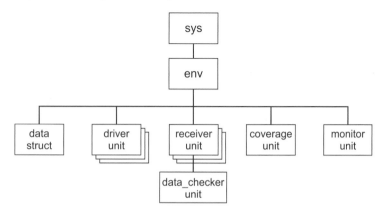

11.2 Execution Flow

One initiates a simulation in Specman Elite by clicking on the *Test* button in the GUI or by issuing the **test** command at the Specman Elite command prompt. To understand how *e* code is executed, it is necessary to understand the phases of execution. This section discusses these execution phases.

Upon issuance of the **test** command, Specman Elite goes through five distinct execution phases. Each test phase is called in a specific order. Each phase is represented by a predefined method in Specman Elite. Each phase is normally empty. User *e* code is executed by extension of some of these phase methods. Figure 11-7 shows these test phases.

Figure 11-7 Execution Test Phases

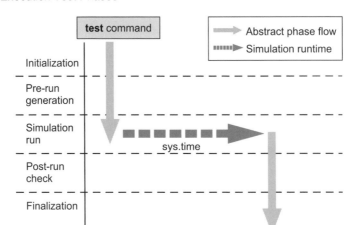

11.2.1 Description of Test Phases

The test phases that are executed by the **test** command are described as follows:

1. **Initialization**: In this phase, Specman Elite does general preparations and configuration. In addition to other methods, the initialization phase calls the **sys.setup()** method. Any *e* simulator configuration options are specified by extension of **sys.setup()** in this phase.

2. **Pre-run Generation**: In this phase, Specman Elite performs memory allocation, instantiation, and generation of the data elements before simulation. In addition to other methods, the pre-run generation phase calls the predefined **init()** and **pre_generate()** methods for every struct instance under **sys**. Then the generation of the data elements occurs. After the generation is complete, the pre-run generation phase calls the **post_generate()** methods for every struct instance under **sys**. Each of these methods can be extended to include method calls. Any computations that are done after Specman Elite has generated the data elements must be done by extension of the **post_generate()** method. Each method call is executed for all instantiated structs before moving to the next method call.

3. **Simulation Run**: The actual simulation occurs in this phase. Any TCMs are executed in this phase. If Specman Elite is working with an HDL simulator, the start of the simulation run is time 0 of the HDL simulator. The predefined **run()** method is called for every struct and unit in the **sys** hierarchy. One can initiate simulation time activity by starting (with the **start** keyword) a TCM from the **run()** method of a struct or unit. For any simulation, at least one main TCM must be started from the **run()** method of a top-level unit. Otherwise, no simulation activity will occur. This main TCM then starts or

calls other TCMs to control execution. Figure 11-8 shows two ways to launch TCMs from the main TCM. Simulation time elapses only in this phase.

Figure 11-8 Two Ways to Launch a TCM from Main TCM

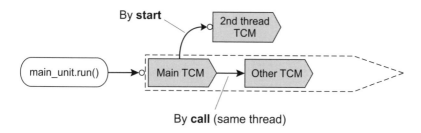

4. **Post-Run Check**: This phase performs preparation for checking and then performs the checking.

5. **Finalization**: This phase does simulation cleanup, coverage, and reporting.

11.2.2 Test Phase Methods

Figure 11-9 shows a detailed flow of the method calls invoked by Specman Elite during each of the five test phases.

Figure 11-9 Test Phase Methods

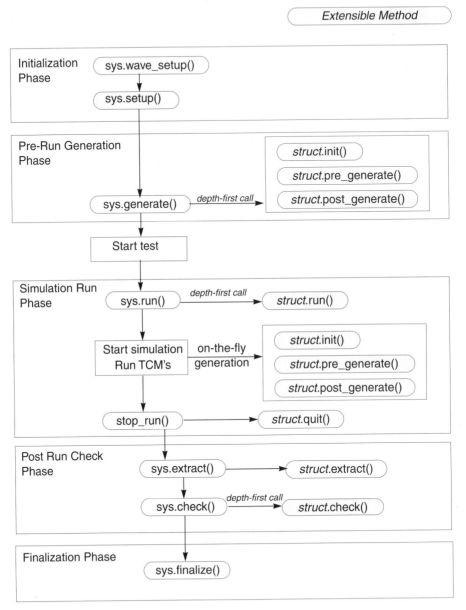

Although there are many extensible methods, the most commonly extended methods are described below.

11.2.2.1 sys.setup()

This method initializes configuration and test parameters for running a test. For example, the seed value for generation is set during this method. The coverage mode is set by extension of this method call. Example 11-1 shows the usage of this method.

Example 11-1 sys.setup() Method

```
Extend the sys.setup() method
<'
extend sys {
    setup() is also {
        set_config(cover, mode, normal);
    };
};
'>
```

11.2.2.2 post_generate()

This method is called for every struct or unit instance in the **sys** hierarchy. The **post_generate()** method is run automatically after an instance of the enclosing struct is allocated and both pre-generation and generation have been performed. You can extend the predefined **post_generate()** method for any struct to manipulate values produced during generation. The **post_generate()** method allows you to derive more complex expressions or values from the generated values.

The order of generation is recursively as follows:

1. Allocate the new struct.

2. Call **pre_generate()**.

3. Perform generation.

4. Call **post_generate()**.

Example 11-2 shows the usage of the **post_generate()** method.

Example 11-2 post_generate() Method

```
Example shows usage of post_generate() method
<'
struct a {
    !m: int; //Do not create m in generate()
    m1: int; //Create a value for m1 in generate()
```

Example 11-2 post_generate() Method (Continued)

```
    post_generate() is also { //Extend the post_generate() method
        m = m1 + 1; //Compute value of m based on generated value of
                    //m1. This method will be called for an instance
                    //of this struct just after the generation is
                    //complete.
    };
};
extend sys {
    A: a; //Instantiate struct a in the sys hierarchy
};
'>
```

11.2.2.3 run()

The **run()** method is extended to start user-defined TCMs. The method is initially empty. The **run()** methods of all structs under **sys** are called, starting from **sys,** in depth-first search order when you execute a **test** or a **run** command. To run the simulation at least one main TCM should be started from the **run()** method of the main unit. Note that **run()** itself is not a TCM. Therefore, you can only **start** TCMs, not call them, from the **run()** method.

Example 11-3 shows the usage of the **run()** method.

Example 11-3 run() Method

```
Example shows usage of run() method
<'
struct packet {
    id : int;
    event clk is rise('~/top/clk') @sim; //Define clk
    monitor() @clk is {
        while (TRUE) {
            wait [2] * cycle;
            out("cycle ", sys.time,
                " Packet id ", id, " is still running");
        };
    };
    run() is also { //Extend the run() method of the struct
        start monitor(); //Start the monitor() TCM from the run()
                         //method of the struct. This TCM will
                         //start at simulation time 0 when the
                         //e simulator hits the simulation run phase.
    };
};
'>
```

11.3 Synchronization between HDL Simulator and Specman Elite

The HDL simulator and Specman Elite are two independent processes running the simulation. Figure 11-10 shows these independent processes that talk to each other through a stub file.

Figure 11-10 Specman Elite and HDL Simulator as Independent Processes

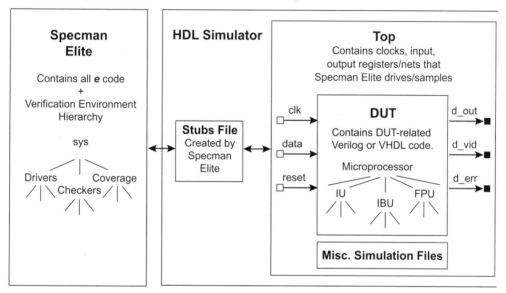

To run the simulation, it is important to understand the order of events and the synchronization between the HDL simulator and Specman Elite. Figure 11-11 shows the synchronizations between the HDL simulator and Specman Elite.

Figure 11-11 Synchronization between Specman Elite and HDL Simulator

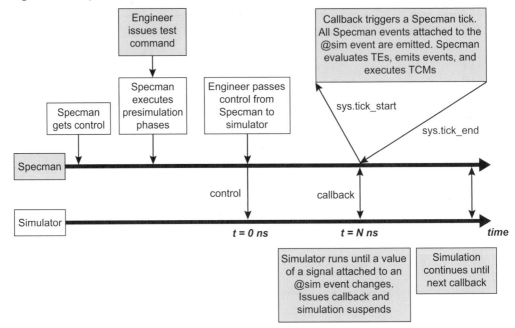

The order of events shown in Figure 11-11 is as follows:

1. Both Specman Elite and the HDL simulator are invoked simultaneously. Specman Elite gets the control of simulation.

2. Verification engineer executes the **test** command.

3. Specman Elite executes the initialization and pre-run generation phases.

4. When Specman Elite gets ready to execute the simulation run phase, the **run()** method for all structs and units in the simulation is called at simulation time t = 0. Any TCMs started in the **run()** method begin execution.

5. Verification engineer passes control of the test from Specman Elite to HDL simulator.

6. Verification engineer starts the simulation in the HDL simulator at simulation time t = 0.

7. The HDL simulator continues simulation until an HDL signal value watched by **@sim** makes the desired transition. At this time, the HDL simulator halts simulation and transfers control to Specman Elite. This is called an HDL simulator callback.

8. All Specman Elite events attached to the **@sim** event are emitted (triggered). Specman Elite evaluates all temporal expressions, emits other events, and executes TCMs. Specman Elite does all this in zero simulation time. At some point, all TCMs are waiting for some simulation time event. Thus the TCMs cannot proceed further without simulation.

At this point, the control is passed back to the HDL simulator. This phase of Specman Elite computation from the time of an HDL simulator callback to the time control is passed back to the HDL simulator is called a *Specman Elite tick*. A Specman Elite tick executes in zero simulation time.

9. HDL simulator regains control. Steps 7 and 8 repeat continuously until a **stop_run()** method call is encountered in the *e* code. This finishes the simulation run phase.

10. Control is transferred back to Specman Elite, which finishes the post-run check and finalization phase to complete the test.

11.4 Summary

- Multiple components make up a verification environment. A typical verification environment contains a data object, a driver object, a receiver object, a data checker object, a monitor object, a coverage object, tests, and an environment object.

- One initiates a test in Specman Elite by clicking on the *Test* button in the GUI or by issuing the **test** command at the Specman Elite command prompt. Upon issuance of the **test** command, Specman Elite goes through five distinct execution phases: initialization, pre-run generation, simulation run, post-run check, and finalization. Each phase is represented by a predefined method in Specman Elite. Each phase is normally empty. User *e* code is executed by extension of some of these phases.

- The most commonly extended test phase methods are **sys.setup()** during the initialization phase, **post_generate()** for every struct or unit instance during the pre-run generation phase, and **run()** for every struct or unit instance during the simulation run phase.

- The HDL simulator and Specman Elite are two independent processes running the simulation. There is a specific synchronization sequence between the HDL simulator and Specman Elite.

11.5 Exercises

1. Explain the components in a typical verification system. Draw a diagram that explains the various components.

2. Describe the five phases during a Specman Elite simulation. Draw a diagram explaining the five test phases.

3. Cite the commonly extended methods in the test phases of a Specman Elite simulation. Give examples of each method.

4. Describe the synchronization between Specman Elite and the HDL simulator. Can simulation time advance in a Specman tick?

PART 3

Creating a Complete Verification System with *e*

CHAPTER 12

Verification Setup and Specification

The previous portions of the book have focused on introductory and syntax aspects of *e*. However, it is important for a verification engineer to understand how to build a complete verification system with *e*. This portion of the book consists of two chapters. This chapter takes the reader through the complete verification process of a simple router design. Topics discussed are design specification, verification components, verification plan, and test plan. The next chapter completes the example with an explanation of the actual *e* code for each component required for the verification of the router design. Note that although a design from the networking domain has been chosen as an example, the principles of verification that are discussed also apply to other domains such as CPU, graphics, video, etc.

At the end of this portion the reader should be able to build a complete verification environment with *e*. The reader should be able to apply methodologies that maximize verification productivity. This section is a great reference for any engineer building a new verification environment or a new component in an existing verification environment.

Chapter Objectives

- Describe the DUT specification.
- Define variable data (**struct**) components.
- Describe static (**unit**) verification components.
- Understand the test scenarios to be executed.

12.1 Device Under Test (DUT) Specification

This section describes the complete DUT specification of a simple router design. This is not as big as a real router but mimics the basic functionality of a real router in a simple manner.

12.1.1 Input/Output Specification

Figure 12-1 shows the input/output specification of the DUT.

Figure 12-1 Input/Output Specification of Router

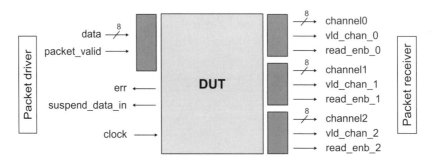

The router accepts data packets on a single 8-bit input port called *data* and routes the packets to one of the three output channels, *channel0*, *channel1*, and *channel2*. In a typical router, packets are accepted from a packet driver ASIC or line card. Similarly, output packets are routed to a packet receiver ASIC or line card. Other I/O signals are described later in this section.

12.1.2 Data Packet Description

A packet is a sequence of bytes with the first byte containing a *header*, the next variable set of bytes containing data, and the last byte containing *parity*. The packet format has the following characteristics.

- The header consists of a 2-bit *address* field and a 6-bit *length* field.
- The *address* field is used to determine to which output channel the packet should be routed, with address "3" being illegal.
- The *length* field specifies the number of data bytes (payload). A packet can have a minimum data size of 1 byte and a maximum data size of 63 bytes.
- The *parity* should be a byte of even, bitwise parity, calculated over the header and data bytes of the packet.

Figure 12-2 shows the format of a data packet. The data packet is modeled as a **struct**.

Figure 12-2 Data Packet Format

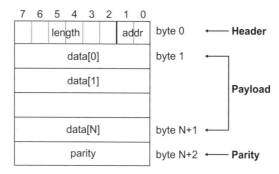

12.1.3 DUT Input Protocol

Figure 12-3 shows the input protocol for the DUT.

Figure 12-3 DUT Input Protocol

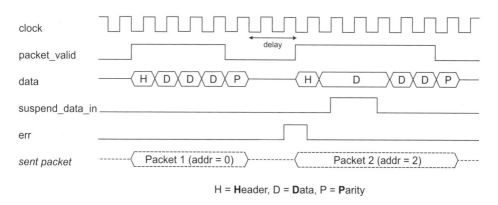

H = **H**eader, D = **D**ata, P = **P**arity

The characteristics of the DUT input protocol are as follows:

- All input signals are active high and are synchronized to the falling edge of the *clock*. Therefore, any signal that is an input to the DUT is driven at the falling edge of *clock*. This is because the DUT router is sensitive to the rising edge of *clock*. Therefore, driving input signals on the falling edge ensures adequate setup and hold time. Since this is a functional simulation rather than a timing simulation, this specification will suffice.

- The *packet_valid* signal has to be asserted on the same *clock* as when the first byte of a packet (the header byte), is driven onto the *data* bus.

- Since the header byte contains the *address*, this tells the router to which output channel the packet should be routed (*channel0*, *channel1*, or *channel2*).
- Each subsequent byte of *data* should be driven on the data bus with each new falling *clock*.
- After the last payload byte has been driven, on the next falling *clock*, the *packet_valid* signal must be deasserted, and the packet parity byte should be driven. This signals packet completion.
- The input *data* bus value cannot change while the *suspend_data_in* signal is active (indicating a FIFO overflow). The packet driver should not send any more bytes and should hold the value on the *data* bus. The width of *suspend_data_in* signal assertion should not exceed 100 cycles.
- The *err* signal asserts when a packet with bad parity is detected in the router, within 1 to 10 cycles of packet completion.

12.1.4 DUT Output Protocol

Figure 12-4 shows the output protocol for the DUT.

Figure 12-4 DUT Output Protocol

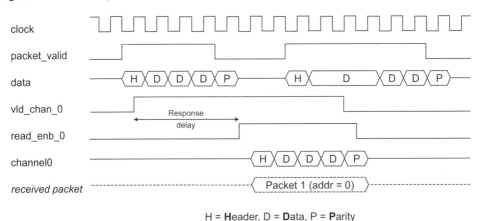

H = **H**eader, D = **D**ata, P = **P**arity

The characteristics of the DUT output protocol are as follows:

- All output signals are active high and are synchronized to the falling edge of the *clock*. Thus, the packet receiver will drive sample data at the falling edge of *clock*. This is correct because the router will drive and sample data at the rising edge of *clock*.
- Each output port *channelX* (*channel0*, *channel1* or *channel2*) is internally buffered by a FIFO of depth 16 and a width of 1 byte.

- The router asserts the *vld_chan_X* (*vld_chan_0, vld_chan_1* or *vld_chan2*) signal when valid data appears on the *channelX* (*channel0, channel1* or *channel2*) output bus. This is a signal to the packet receiver that valid data is available on a particular router.

- The packet receiver will then wait until it has enough space to hold the bytes of the packet and then respond with the assertion of the *read_enb_X* (*read_enb_0, read_enb_1* or *read_enb_2*) signal that is an input to the router.

- The *read_enb_X* (*read_enb_0, read_enb_1* or *read_enb_2*) input signal is asserted on the falling *clock* edge in which data are read from the *channelX* (*channel0, channel1* or *channel2*) bus.

- As long the *read_enb_X* (*read_enb_0, read_enb_1* or *read_enb_2*) signal remains active, the *channelX* (*channel0, channel1* or *channel2*) bus drives a valid packet byte on each rising *clock* edge.

- The packet receiver cannot request the router to suspend data transmission in the middle of a packet. Therefore, the packet receiver must assert the *read_enb_X* (*read_enb_0, read_enb_1* or *read_enb_2*) signal only after it ensures that there is adequate space to hold the entire packet.

- The *read_enb_X* (*read_enb_0, read_enb_1* or *read_enb_2*) must be asserted within 30 *clock* cycles of the *vld_chan_X* (*vld_chan_0, vld_chan_1* or *vld_chan2*) being asserted. Otherwise, there is too much congestion in the packet receiver.

- The DUT *channelX* (*channel0, channel1* or *channel2*) bus must not be tri-stated (high-Z) when the DUT signal *vld_chan_X* (*vld_chan_0, vld_chan_1* or *vld_chan2*) is asserted (high) and the input signal *read_enb_X* (*read_enb_0, read_enb_1* or *read_enb_2*) is also asserted high.

12.1.5 DUT State Machine

Figure 12-5 shows the state machine transitions for the DUT. All state transitions in the DUT occur at the rising edge of *clock*.

Figure 12-5 DUT State Machine

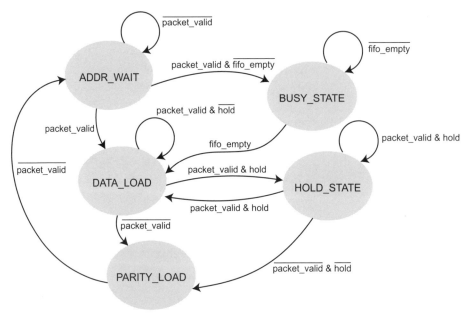

The legal transitions of the DUT state machine shown in Figure 12-5 above are listed in Figure 12-6. All transitions that are not listed below are illegal transitions. It is important to monitor and report an error message when such transitions happen.

Figure 12-6 DUT State Machine Transitions

Current State	Next State
ADDR_WAIT = 0 x 0	ADDR_WAIT, DATA_LOAD, BUSY _STATE
DATA_LOAD = 0 x 1	DATA_LOAD, PARITY_LOAD, HOLD_STATE
PARITY_LOAD = 0 x 2	ADDR_WAIT
HOLD_STATE = 0 x 3	HOLD_STATE, DATA_LOAD, PARITY_LOAD
BUSY_STATE = 0 x 4	BUSY_STATE, DATA_LOAD

12.2 DUT HDL Source Code

Example 12-1 shows the Verilog source code used to describe the DUT router.

Example 12-1 DUT HDL Source Code

```
//*****************************************************************/
//Source code for the Router DUT
//Contains the following modules
//top: top level module that instantiates the router
//router: router DUT instantiates three fifo modules +
//        port_fsm(state machine)
//port_fsm: Implements the router state machine.
//fifo: Implements the fifo that holds data for each channel
//*****************************************************************/
//*****************************************************************/
//Source code for fifo module
module fifo (clock,
             write_enb,
             read_enb,
             data_in,
             data_out,
             empty,
             full);
input      clock;
input write_enb;
input read_enb;
input  [7:0] data_in;
output [7:0] data_out;
output empty;
output full;
```

Example 12-1 DUT HDL Source Code (Continued)

```verilog
// Port Signals
wire      clock;
wire write_enb;
wire read_enb;
wire   [7:0] data_in;
reg    [7:0] data_out;
wire empty;
wire full;

// Internal Signals
reg       [7:0] ram[0:15];
reg             tmp_empty;
reg             tmp_full;
integer         write_ptr;
integer         read_ptr;

// Processes
   initial begin
      data_out  = 8'b0000_0000;
      tmp_empty = 1'b1;
      tmp_full  = 1'b0;
      write_ptr = 0;
      read_ptr  = 0;
   end

   assign empty = tmp_empty;
   assign full  = tmp_full;
 always @(posedge clock) begin : fifo_core
      if ((write_enb == 1'b1) &&  (tmp_full == 1'b0)) begin
         ram[write_ptr] = data_in;
         tmp_empty <= 1'b0;
         write_ptr = (write_ptr + 1) % 16;
         if ( read_ptr == write_ptr ) begin
            tmp_full <= 1'b1;
         end //if
      end //if
```

Example 12-1 DUT HDL Source Code (Continued)

```
 if ((read_enb == 1'b1) &&  (tmp_empty == 1'b0)) begin
         data_out <= ram[read_ptr];
         tmp_full <= 1'b0;
         read_ptr = (read_ptr + 1) % 16;
         if ( read_ptr == write_ptr ) begin
            tmp_empty <= 1'b1;
         end //if
      end //if
   end  //fifo_core;
endmodule //fifo


//************************************************************/
//Source code for port_fsm module (DUT State Machine)
`define ADDR_WAIT     4'b0000
`define DATA_LOAD     4'b0001
`define PARITY_LOAD   4'b0010
`define HOLD_STATE    4'b0011
`define BUSY_STATE    4'b0100
module port_fsm (clock,
                 suspend_data_in,
                 err,
                 write_enb,
                 fifo_empty,
                 hold,
                 packet_valid,
                 data_in,
                 data_out,
                 addr);
input       clock;
output      suspend_data_in;
output      err;
output[2:0]  write_enb;
input   fifo_empty;
input       hold;
input       packet_valid;
input[7:0]  data_in;
output[7:0]  data_out;
output   [1:0]      addr;
```

Example 12-1 DUT HDL Source Code (Continued)

```verilog
reg [7:0]  data_out;
reg [1:0]  addr;
// Internal Signals
reg    [2:0] write_enb_r;
reg          fsm_write_enb;
reg    [3:0] state_r;
reg    [3:0] state;
reg    [7:0] parity;
reg    [7:0] parity_delayed;
reg          sus_data_in;
// Processes
  initial begin
       err            = 1'b0;
       data_out       = 8'b0000_0000;
       addr           = 2'b00;
       write_enb_r    = 3'b000;
       fsm_write_enb  = 1'b0;
       state_r        = 4'b0000;
       state          = 4'b0000;
       parity         = 8'b0000_0000;
       parity_delayed = 8'b0000_0000;
       sus_data_in    = 1'b0;
  end
  assign suspend_data_in = sus_data_in;
  always @(packet_valid) begin : addr_mux
    if (packet_valid == 1'b1) begin
      case (data_in[1:0])
      2'b00 :  begin
             write_enb_r[0] = 1'b1;
             write_enb_r[1] = 1'b0;
             write_enb_r[2] = 1'b0;
      end
      2'b01 :  begin
        write_enb_r[0] = 1'b0;
        write_enb_r[1] = 1'b1;
        write_enb_r[2] = 1'b0;
       end
      2'b10 :  begin
        write_enb_r[0] = 1'b0;
        write_enb_r[1] = 1'b0;
        write_enb_r[2] = 1'b1;
      end
      default :write_enb_r = 3'b000;
    endcase
    end //if
end //addr_mux;
```

Example 12-1 DUT HDL Source Code (Continued)

```verilog
always @(posedge clock) begin : fsm_state
    state_r <= state;
end //fsm_state;

always @(state_r or packet_valid or fifo_empty or hold or data_in)
begin : fsm_core
state = state_r;   //Default state assignment
    case (state_r)
       `ADDR_WAIT :   begin
                //transition//
                if ((packet_valid == 1'b1) &&
                    (2'b00 <= data_in[1:0]) &&
                    (data_in[1:0] <= 2'b10)) begin
                    if (fifo_empty == 1'b1) begin
                      state = `DATA_LOAD;
                    end
                    else begin
                      state = `BUSY_STATE;
                    end //if
                 end //if;
                //combinational//
                sus_data_in = !fifo_empty;
                if ((packet_valid == 1'b1) &&
                    (2'b00 <= data_in[1:0]) &&
                    (data_in[1:0] <= 2'b10) &&
                    (fifo_empty == 1'b1)) begin
                        addr = data_in[1:0];
                        data_out  = data_in;
                        fsm_write_enb = 1'b1;
                end
                else begin
                    fsm_write_enb = 1'b0;
                end //if
              end // of case ADDR_WAIT
        `PARITY_LOAD : begin
                //transition//
                state = `ADDR_WAIT;
                //combinational//
                data_out = data_in;
                fsm_write_enb = 1'b0;
              end // of case PARITY_LOAD
        `DATA_LOAD :   begin
             //transition//
             if ((packet_valid == 1'b1) &&
               (hold == 1'b0)) begin
                  state = `DATA_LOAD
             end
```

Example 12-1 DUT HDL Source Code (Continued)

```
              else if ((packet_valid == 1'b0) &&
              (hold == 1'b0)) begin
                  state = `PARITY_LOAD;
              end
              else begin
                  state = `HOLD_STATE;
              end  //if
              //combinational//
              sus_data_in = 1'b0;
              if ((packet_valid == 1'b1) &&
                (hold == 1'b0)) begin
                  data_out = data_in;
                  fsm_write_enb = 1'b1;
              end
              else if ((packet_valid == 1'b0) &&
                (hold == 1'b0)) begin
                  data_out = data_in;
                  fsm_write_enb = 1'b1;
              end
              else begin
              fsm_write_enb = 1'b0;
              end //if
          end  //end of case DATA_LOAD
      `HOLD_STATE :  begin
                  //transition//
              if (hold == 1'b1) begin
                  state = `HOLD_STATE;
              end
              else if ((hold == 1'b0) && (packet_valid == 1'b0)) begin
                  state = `PARITY_LOAD;
              end
              else begin
                  state = `DATA_LOAD;
              end //if
                  //combinational
              if (hold == 1'b1) begin
                  sus_data_in = 1'b1;
                  fsm_write_enb = 1'b0
              end
              else begin
                  fsm_write_enb = 1'b1;
                  data_out = data_in;
              end //if
          end  //end of case `HOLD_STATE
```

Example 12-1 DUT HDL Source Code (Continued)

```
              `BUSY_STATE :  begin
              //transition//
              if (fifo_empty == 1'b0) begin
                    state = `BUSY_STATE;
              end
              else begin
                    state = `DATA_LOAD;
              end //if
              //combinational//
              if (fifo_empty == 1'b0) begin
                    sus_data_in = 1'b1;
              end
              else begin
                    addr = data_in[1:0]; // hans
                    data_out  = data_in;
                    fsm_write_enb = 1'b1;
              end //if
        end  //end of case BUSY_STATE
  endcase
end //fsm_core

assign write_enb[0] = write_enb_r[0] & fsm_write_enb;
assign write_enb[1] = write_enb_r[1] & fsm_write_enb;
assign write_enb[2] = write_enb_r[2] & fsm_write_enb;

always @(posedge clock) begin : parity_calc
    parity_delayed <= parity;
    err <= 1'b0;
    if ((packet_valid == 1'b1) && (sus_data_in == 1'b0)) begin
      parity <= parity ^ data_in;
    end
    else if (packet_valid == 1'b0) begin
      if ((state_r == `PARITY_LOAD) && (parity_delayed != data_in))
      begin
        err <= 1'b1;
      end
      else begin
        err <= 1'b0;
      end //if
    parity <= 8'b0000_0000;
    end //if
  end //parity_calc;
endmodule //port_fsm
//********************************************************************/
```

Example 12-1 DUT HDL Source Code (Continued)

```
//******************************************************************/
//Source code for module router, instantiates 3 fifo modules +
//port_fsm (State machine for router)
module router (clock,
               packet_valid,
               data,
               channel0,
               channel1,
               channel2,
               vld_chan_0,
               vld_chan_1,
               vld_chan_2,
               read_enb_0,
               read_enb_1,
               read_enb_2,
               suspend_data_in,
               err);
input          clock;
input          packet_valid;
input    [7:0] data;
output   [7:0] channel0;
output   [7:0] channel1;
output   [7:0] channel2;
output         vld_chan_0;
output        vld_chan_1;
output        vld_chan_2;
input         read_enb_0;
input         read_enb_1;
input         read_enb_2;
output         suspend_data_in;
output         err;
// Internal Signals
wire   [7:0] data_out_0;
wire   [7:0] data_out_1;
wire   [7:0] data_out_2;
wire full_0;
wire full_1;
wire full_2;
wire empty_0;
wire empty_1;
wire empty_2;
wire fifo_empty;
wire fifo_empty0;
wire fifo_empty1;
wire fifo_empty2;
```

Example 12-1 DUT HDL Source Code (Continued)

```
wire hold_0;
wire hold_1;
wire hold_2;
wire hold;
wire   [2:0] write_enb;
wire   [7:0] data_out_fsm;
wire   [1:0] addr;

// Instantiations
  fifo queue_0 (.clock     (clock),
               .write_enb (write_enb[0]),
               .read_enb  (read_enb_0),
               .data_in   (data_out_fsm),
               .data_out  (data_out_0),
               .empty     (empty_0),
               .full      (full_0));

  fifo queue_1 (.clock     (clock),
               .write_enb (write_enb[1]),
               .read_enb  (read_enb_1),
               .data_in   (data_out_fsm),
               .data_out  (data_out_1),
               .empty     (empty_1),
               .full      (full_1));

  fifo queue_2 (.clock     (clock),
               .write_enb (write_enb[2]),
               .read_enb  (read_enb_2),
               .data_in   (data_out_fsm),
               .data_out  (data_out_2),
               .empty     (empty_2),
               .full      (full_2));

  port_fsm in_port (.clock          (clock),
                   .suspend_data_in (suspend_data_in),
                   .err             (err),
                   .write_enb       (write_enb),
                   .fifo_empty      (fifo_empty),
                   .hold            (hold),
                   .packet_valid    (packet_valid),
                   .data_in         (data),
                   .data_out        (data_out_fsm),

                   .addr            (addr));
```

Example 12-1 DUT HDL Source Code (Continued)

```
// Processes
  assign channel0 = data_out_0;   //make note assignment only for
                                  //consistency with vlog env
  assign channel1 = data_out_1;
  assign channel2 = data_out_2;

  assign vld_chan_0 = !empty_0;
  assign vld_chan_1 = !empty_1;
  assign vld_chan_2 = !empty_2;

  assign fifo_empty0 = (empty_0 | ( addr[1] |  addr[0])); //addr!=00
  assign fifo_empty1 = (empty_1 | ( addr[1] | !addr[0])); //addr!=01
  assign fifo_empty2 = (empty_2 | (!addr[1] |  addr[0])); //addr!=10

  assign fifo_empty  = fifo_empty0 & fifo_empty1 & fifo_empty2;

  assign hold_0 = (full_0 & (!addr[1] & !addr[0])); //addr=00
  assign hold_1 = (full_1 & (!addr[1] &  addr[0])); //addr=01
  assign hold_2 = (full_2 & ( addr[1] & !addr[0])); //addr=10
  assign hold   = hold_0 | hold_1 | hold_2;
endmodule //router
//*******************************************************************/
//Source code for top level module
//Instantiates the router module
module top();
  reg          clock;
  reg          packet_valid;
  reg    [7:0] data;
  wire   [7:0] channel0;
  wire   [7:0] channel1;
  wire   [7:0] channel2;
  wire         vld_chan_0;
  wire         vld_chan_1;
  wire         vld_chan_2;
  reg          read_enb_0;
  reg          read_enb_1;
  reg          read_enb_2;
  wire         suspend_data_in;
  wire         err;

  router router1 (.clock          (clock),
                  .packet_valid   (packet_valid),
                  .data           (data),
                  .channel0       (channel0),
```

Example 12-1 DUT HDL Source Code (Continued)

```
                        .channel1          (channel1),
                        .channel2          (channel2),
                        .vld_chan_0        (vld_chan_0),
                        .vld_chan_1        (vld_chan_1),
                        .vld_chan_2        (vld_chan_2),
                        .read_enb_0        (read_enb_0),
                        .read_enb_1        (read_enb_1),
                         .read_enb_2        (read_enb_2),
                        .suspend_data_in(suspend_data_in),
                        .err               (err));
// Processes
   initial begin
      packet_valid = 1'b0;
      data         = 8'b0000_0000;
      read_enb_0   = 1'b0;
      read_enb_1   = 1'b0;
      read_enb_2   = 1'b0;
   end

   initial begin
      clock = 0;
      forever begin
         #50 clock = !clock;
      end
   end
endmodule //top
```

12.3 Verification Plan

A verification plan is required to describe what is to be verified and how it will be verified. It should address the three aspects of verification: coverage measurement, stimulus generation, and response checking. The verification environment will be developed in *e*. Multiple components are needed in the router verification environment. Figure 12-7 shows the components in the router verification environment.

Figure 12-7 Components of the Router Verification Environment

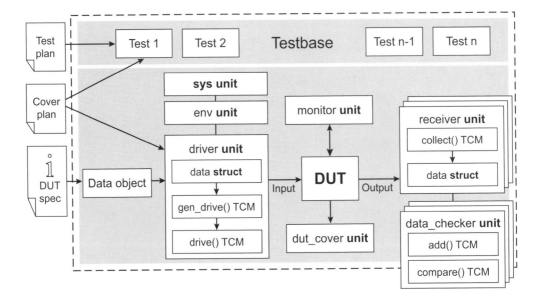

A summary description of the objects follows—

12.3.1 Data Object (Stimulus Generation)

The generation component of verification uses *data objects* for stimulus generation. These data objects are structs that represent the abstract form of input stimulus. In the router verification, the data object is in form of the **struct** *packet*. This struct is the basic form of stimulus.

12.3.2 Driver Object (Stimulus Driver)

The *driver* object performs the function of taking the stimulus data objects one at a time and applying them to the DUT until all stimulus has been applied. Typically, the driver object is similar to an HDL bus functional model. In case of a networking environment, it is also known as a *port* object. In the router environment the driver object is called *sbt_driver* and is modeled as a **unit**.

12.3.3 Receiver Object (Output Receiver)

The *receiver* object is responsible for collecting the data at output of the DUT. The *receiver* object performs the function of taking the raw output data from the DUT based on the output protocol and converting the data to an abstract object for comparison. Typically, the receiver object is similar to an HDL bus functional model. In the router environment the receiver object is called *sbt_receiver* and is modeled as a **unit**. There are three instances of the *sbt_receiver*, one per channel.

12.3.4 Data Checker Object (Expected Value Comparison)

The *data_checker* object stores expected values for each input data object injected into the DUT. When the receiver object receives one data object, it passes it to the data_checker for comparison against the expected values. In the router environment, the data_checker object will be modeled as the *sbt_scoreboard* **unit**.

12.3.5 Monitor Object (Protocol Checker)

A *monitor* object checks the input and output timing protocol. In the router environment, a monitor object is modeled as the *sbt_checker* **unit**. Various protocol checks are included in this unit.

12.3.6 Coverage Object (DUT and *e* Coverage)

A *coverage* object sets up coverage tracking on key items in the *e* code and in the DUT. Some coverage objects can also be embedded in the extension to a driver or receiver object. In the router environment, the coverage object is modeled as *sbt_dut_cover* **unit**.

12.3.7 Env Object (Environment)

All verification components in this section are instantiated under the *sbt_env* **unit**. The *sbt_env* **unit** is instantiated in **sys**. One can transport the *sbt_env* **unit** to any other level of hierarchy without having to transport each individual verification component.

12.4 Test Plan

Tests are derived from the test plan. A test plan contains all tests that are to be run to verify the DUT. Typically, a test plan contains an exhaustive list of items to be tested. In *e*, tests are simply extensions of existing struct and unit definitions that impose additional constraints on these objects. Test files are very small and are easy to write and to maintain.

In this book, we will look at two simple test scenarios as a part of the test plan. Details of these test scenarios will be discussed in the following chapter.

12.4.1 Test Scenario 1 (Distribution of Packets)

Create packets with a certain probability distribution. There should be a mix of packets of different lengths. There should also be a mix of good and bad parity packets.

12.4.2 Test Scenario 2 (Protocol Violation)

Create a test that will make the packet receiver assert the *read_enb_X* (*read_enb_0, read_enb_1* or *read_enb_2*) more than 30 *clock* cycles after *vld_chan_X* (*vld_chan_0, vld_chan_1* or *vld_chan2*) is asserted. This will violate the protocol.

12.5 Summary

- This chapter describes the complete DUT specification of a simple router design. This is not as big as a real router but mimics the basic functionality of a real router in a simple manner.

- Input/output specification, packet format, input protocol, output protocol, and the state machine description were discussed for the DUT.

- The HDL source code for the router design was discussed. The router is built from FIFOs and a state machine module. A top-level module instantiates the router module.

- The verification environment consists of data object (**struct** *packet*), driver object (**unit** *sbt_driver*), receiver object (**unit** *sbt_receiver*), data checker object (**unit** *sbt_scoreboard*), monitor object (**unit** *sbt_checker*), coverage object (**unit** *sbt_dut_cover*), and environment object (**unit** *sbt_env*).

- A typical test plan covers all possible scenarios for which a DUT needs to be tested. In this chapter, we have only described two simple test scenarios.

CHAPTER 13

Creating and Running the Verification Environment

The previous chapter takes the reader through the complete verification process of a simple router design. Topics discussed are design specification, verification components, verification plan, and test plan. This chapter completes the example with an explanation of the actual *e* code for each component required for the verification of the router design. Note that although a design from the networking domain has been chosen as an example, the principles of verification that are discussed also apply to other domains such as CPU, graphics, video, etc.

At the end of this chapter the reader should understand how to build a complete verification environment with *e*. The reader should be able to apply methodologies that maximize verification productivity.

Chapter Objectives

- Describe the *e* code for variable data (**struct**) components.
- Understand the *e* code for static (**unit**) verification components.
- Write the *e* code for the test scenarios to be executed.
- Explain how to put the *e* code together to run the simulation.

13.1 Defining the Packet Data Item

This section shows the requirements and *e* code for the packet data item.

13.1.1 Requirements for the Packet Data Item

The specification for the packet data items is described in "Data Packet Description" on page 224. Requirements for the packet data item are as follows:

- Since the packet data item is expected to be generated on the fly, it should be defined as a **struct**.

- Define an enumerated type to describe the payload. The values for that enumerated type are SMALL, MEDIUM, and LARGE. Create a field *payload_size* of that enumerated type.

- Define an 8-bit field *uid*.

- The field *addr* is two bits wide and its values should be constrained between 0 and 2 because there are only three output channels. This should be a soft constraint as another channel may be added later.

- When *payload_size* is SMALL, the *len* field is constrained between [1..20]; when *payload_size* is MEDIUM, the len field is constrained between [21..44]; and when *payload_size* is LARGE, the len field is constrained between [45..63].

- The field *data* item is a list of bytes. The length of the *data* list is equal to the *len* field.

- The first byte of the *data* list is equal to the *uid* field.

- Define an enumerated type to describe the type of parity. The values for that enumerated type are GOOD and BAD. Create a field *packet_kind* of that enumerated type.

- When *packet_kind* is GOOD, the *parity* field is constrained to the correct value of parity, and when *packet_kind* is BAD, the *parity* field is constrained to the incorrect value of parity.

- Parity is computed by means of a *parity_calc()* method.

13.1.2 *e* Code for the Packet Data Item

Example 13-1 presents an *e* code for the packet data item.

Example 13-1 *e* Code for the Packet Data Item

```
Definition  :    Fields and constraints defining the packet
<'
-- Define the enumerated types
type sbt_payload_size_t : [SMALL, MEDIUM, LARGE];
type sbt_packet_kind_t  : [GOOD, BAD];


struct sbt_packet {
    payload_size : sbt_payload_size_t;   -- Control field for the
                                         -- payload size
    uid        : byte;                   -- Unique id field
  %addr        : uint(bits : 2);
      keep soft addr in [0..2];    -- Only 3 port addresses, soft in
                                   -- case need to change in future


    %len         : uint(bits : 6);
        -- Constrain the len according the payload_size control
        -- field
        keep payload_size == SMALL  => len in [1..20];
        keep payload_size == MEDIUM => len in [21..44];
        keep payload_size == LARGE  => len in [45..63];

    %data[len]       : list of byte; -- Constrain length of list == len
        -- Constrain the first data byte to be equal to the uid
        keep data[0] == uid;

    packet_kind : sbt_packet_kind_t;    -- Control field for GOOD/BAD
                                        -- parity
        keep soft packet_kind == GOOD;  -- Default: GOOD

    parity_computed : byte; -- Correct value of parity
        -- Assign the calculated parity to parity_computed
        keep parity_computed == parity_calc(len, addr, data);
    %parity      : byte; -- May be assigned either a good or bad value

    -- Assign the correct parity value to the parity field
    when GOOD sbt_packet {
        keep parity == parity_computed;
    };
```

Example 13-1 *e* Code for the Packet Data Item (Continued)

```
-- Do NOT assign the correct parity value to the parity field
-- (random)
when BAD sbt_packet {
    keep parity != parity_computed;
};
-----------------------------------------------------------------
-- parity_calc()
--
-- Return the byte resulting from xor-ing among all data bytes
-- and the byte resulting from concatenating addr and len
-----------------------------------------------------------------
parity_calc( len : uint(bits : 6),
             addr: uint(bits : 2),
             data: list of byte): byte is {
    result = %{len, addr}; -- Concatenation
    for each (data_byte) in data do {    -- Iterate through the
                                         -- data list
       result ^= data_byte ; -- Same as:
                             --result = result^data_byte;
    };
}; -- end parity_calc

}; -- end sbt_packet
'>
```

13.2 Driver Object

This section shows the requirements and *e* code for the driver object.

13.2.1 Requirements for the Driver Object

The driver object (*sbt_driver*) follows the specification described in "DUT Input Protocol" on page 225. Requirements for the driver object are as follows:

- Since the driver object is expected to be static, it should be defined as a **unit**.

- Define a field *cur_packet* of the *sbt_packet* type that holds the values for the current packet being injected into the DUT. Do not generate the field in the generation phase.

- Define a field *delay* of type **uint** that specifies the number of cycles to delay between consecutive packets. Set a soft constraint for the *delay* field to be between 1 and 10.

- Define a field *no_of_pkts* of type **uint** that specifies the number of packets to be injected. Set a soft constraint for the *no_of_pkts* field to be between 5 and 20.

- Define an event *clk_fall* that is emitted at the falling edge of the DUT clock.

- Define a *packet_started* event that is emitted explicitly by the driver object at the beginning of packet transmission.

- Define a *packet_ended* event that is emitted explicitly by the driver object at the end of packet transmission.

- The method *gen_and_drive()* is the master TCM that generates many packets and injects them into the DUT. While generating packets, *uid* field should contain the serial number of the packet being generated.

- The method *drive_packet()* injects exactly one packet into the DUT according to "DUT Input Protocol" on page 225.

Example 13-2, starting on the next page, presents *e* code for the driver object.

13.2.2 *e* Code for the Driver Object

Example 13-2 *e* Code for the Driver Object

```
File contains e code for the driver object.
Since this object is modeled as a unit, all references
to HDL signals assume that an hdl_path() has been defined
at a higher level.

<'
import sbt_packet;

unit sbt_driver { -- Define the driver as a unit

    !cur_packet    : sbt_packet;         -- Packet that will be
                                         -- generated
                                         -- on the fly
    !sent_packets : list of sbt_packet;-- History list of all packets
                                         -- Good debug aid
    !delay        : uint;                -- Generated on the fly
                                         -- for each packet
        keep soft delay in [1..10];      -- Default range
    no_of_pkts : uint; -- Field to control number of packets generated
        keep soft no_of_pkts in [5..20]; -- Soft constraint

    event clk_fall is fall('clock')@sim; -- Clock event

    event packet_started; -- Emitted manually at start of packet
    event packet_ended; -- Emitted manually at end of packet

    -- Event to check that FIFOs for router are not full
    event suspend_data_off is true('suspend_data_in' == 0)@clk_fall;

    -- gen_and_drive()
    --
    -- Drive packets to the DUT input port.
    -- Packets are generated on the fly, each packet generated
    -- just before being sent.

    gen_and_drive() @clk_fall is {
        -- Loop through the number of packets to be generated
        for i from 0 to no_of_pkts - 1 {
            -- Generate and wait a delay before driving the packet
            gen delay;
            wait [delay];
```

Example 13-2 *e* Code for the Driver Object (Continued)

```
              -- Generate the packet keeping the uid equal to the index
              gen cur_packet keeping {
                  .uid == i;
              };

              -- Pass the generated packet to the drive_packet() TCM
              drive_packet(cur_packet);
          };

      wait [500]; -- Arbitrary amount of wait to flush out packets
      stop_run();      -- Stop the simulation

  }; -- end gen_and_drive()
  drive_packet(packet : sbt_packet)  @clk_fall is {
      var pkt_packed: list of byte; -- List of bytes that will be
                                    -- driven to the DUT

      -- Break the packet down into a list of bytes
      -- Hint: Make sure the relevant fields are marked as physical
      pkt_packed = pack(packing.low, packet);

      emit packet_started;
      out(sys.time, " : Starting to drive packet to port ",
          packet.addr);

      -- Start sending
      -- Loop through the list of bytes and drive each byte to the
      -- data signal
      for each (pkt_byte) in pkt_packed {

          while 'suspend_data_in' == 1 {
              wait cycle;
          };
          -- Alternate (elegant) solution of the while loop above:
          -- sync @suspend_data_off;

          -- Drive the packet_valid signal to '1'
          'packet_valid' = 1 ;

          'data[7:0]' =  pkt_byte;         -- Current byte in
                                           -- pkt_packed
```

Example 13-2 *e* Code for the Driver Object (Continued)

```
                -- Drive the packet_valid signal back to 0
                -- after the last data byte (payload) was driven and
                -- before the parity byte is driven (according to the
                -- protocol spec)
                if (index == (pkt_packed.size() - 1)) {
                    'packet_valid' = 0;
                };

                -- Wait a cycle
                wait [1];

        }; -- end for each


 emit packet_ended;

            -- Add the sent packet to the history list
            sent_packets.add(cur_packet);

        }; -- end drive_packet()

        run() is also {
            start gen_and_drive();       -- Start the main TCM at
                                         -- the start of simulation
        };

}; -- end struct sbt_driver

'>
```

13.3 Receiver Object

This section discusses the requirements and *e* code for the receiver object.

13.3.1 Requirements for the Receiver Object

The receiver object (*sbt_receiver*) follows the specification described in "DUT Output Protocol" on page 226. Requirements for the receiver object are as follows:

- Since the receiver object is expected to be static, it should be defined as a **unit**. There will be three instances of the receiver object, each with a different **hdl_path()** corresponding to each channel.

- Define a field *port_no* of type **uint** that specifies the channel with which that receiver object is associated.

- Define a field *rcv_delay* of type **uint** that specifies the number of cycles to delay the assertion of *vld_chan_X* (*vld_chan_0*, *vld_chan_1* or *vld_chan2*) by the router to the assertion of *read_enb_X* (*read_enb_0*, *read_enb_1* or *read_enb_2*) by the receiver object. Set a soft constraint on *rcv_delay* to be equal to 10 cycles.

- Define an event *clk_fall* that is emitted at the falling edge of the DUT clock.

- Define a *packet_start* event that is emitted when the *vld_chan_X* (*vld_chan_0*, *vld_chan_1* or *vld_chan2*) is asserted by the router.

- Define a *pkt_received* event that is emitted when the receiver object completes the reception of one complete packet.

- The method *collect_packets()* sits in a continuous loop waiting for packets from the DUT according to "DUT Output Protocol" on page 226.

13.3.2 *e* Code for the Receiver Object

Example 13-3 presents *e* code for the receiver object.

Example 13-3 *e* Code for the Receiver Object

```
File contains e code for the receiver object.
Since this object is modeled as a unit, all references
to HDL signals assume that an hdl_path() has been defined
at a higher level.
<'
-- Import the sbt_packet
import sbt_packet;

unit sbt_receiver {

    port_no : uint;                 -- port_no of corresponding port

    !rcv_packet : sbt_packet;    -- Packet that will be received

    rcv_delay : uint;               -- Delay that the receiver waits
                                    -- before setting the read enable
        keep soft rcv_delay == 10; -- Soft constraint

    event clk_fall is fall('clock')@sim; -- Synchronizing clock

    -- Event telling the receiver that valid data are available for
    -- reception
    event packet_start is rise('vld_chan_(port_no)')@clk_fall;
```

Example 13-3 *e* Code for the Receiver Object (Continued)

```
-- Event that will be emitted when a packet is fully received
event pkt_received;

-- collect_packets()
--
-- Collect data from output port #<port_no>
collect_packets() @clk_fall is {

    -- List of bytes that will hold the received bytes
    -- from the channelx port
    var received_bytes : list of byte;

    -- Loop forever
    while TRUE {

        -- Wait until valid data are available
        wait until @packet_start;

        -- Wait a delay until receiving
        wait [rcv_delay];

        -- Set the read enable signal
        'read_enb_(port_no)' = 1;

        -- While data are valid on the channel loop
        while 'vld_chan_(port_no)' == 1 {

            -- Wait a cycle
            wait [1];

            -- Add the received byte to the list of bytes
            received_bytes.add('channel(port_no)');
            out(sys.time, " receiving data from channel ",
                port_no);
        };
```

Example 13-3 *e* Code for the Receiver Object (Continued)

```
                -- Release the read enable signal
                'read_enb_(port_no)' = 0;

                -- Unpack the list of bytes to a sbt_packet struct
                -- (rcv_packet)
                unpack(packing.low, received_bytes, rcv_packet);

                -- Delete the list of bytes after the packet is received
                received_bytes.clear();

                -- Emit the pkt_received event
                emit pkt_received;
            };
        };
    -- Start the collect_packets() TCM
        run() is also {
            start collect_packets();
        };
    }; -- end sbt_receiver
    '>
```

13.4 Data Checker Object

This section discusses the requirements and *e* code for the data checker object.

13.4.1 Requirements for the Data Checker Object

The data checker object is implemented in form of a scoreboard (*sbt_scoreboard*). Requirements for the scoreboard are as follows:

- Since the scoreboard is expected to be static, it should be defined as a **unit**. The scoreboard is instantiated inside the receiver object because there needs to be separate data checking per channel.

- Define a list *exp_packets* of type *sbt_packet* that holds the list of expected packets on that channel. Do not generate this list during the generation phase.

- Define a method *add_packet( )* that takes one packet as an argument and adds it to the list of expected packets.

- Define a method *check_packet( )* that takes one packet as an argument and checks it for equality against the first element of the *exp_packets* list.

- Extend the *sbt_driver* defined in "Driver Object" on page 246. At every *packet_started* event, add a packet to the *exp_packets* list.

- Extend the *sbt_receiver* defined in "Receiver Object" on page 250. At every *pkt_received* event, check the received packet against the first element in the *exp_packets* list.

13.4.2 *e* Code for the Data Checker (Scoreboard) Object

Example 13-4 presents *e* code for the data checker (scoreboard) object.

Example 13-4 *e* Code for the Data Checker (Scoreboard) Object

```
File contains e code for the scoreboard object.
This object is instantiated in the receiver object.
<'
import sbt_packet;

unit sbt_scoreboard {

    -- Expected list of packets that have been driven into the DUT
    !exp_packets : list of sbt_packet;

    -- add_packet()
    -- Adds a new packet to the exp_packets list
    add_packet(packet_in : sbt_packet) is {
        exp_packets.add(packet_in);
    };
    -- check_packet()
    -- Tries to find the received packet in the exp_packets list.
    -- Since the order of packets that have been sent into the DUT is
    -- the same for the outcoming packets on the receiver side,
    -- the first packet in the exp_packets list
    -- has to match the received packet. Each time we check a packet
    -- we delete the matched (first) packet from the expected packets
    -- list.
    check_packet(packet_out : sbt_packet) is {
        var diff : list of string;

        -- Compare the physical fields (addr, len, data, parity) of
        -- the received packet with the first packet in the
        -- exp_packets list.
        -- The last parameter indicates that we only care to report up
        -- to 10 differences.
        diff = deep_compare_physical(exp_packets[0], packet_out, 10);
```

Example 13-4 *e* Code for the Data Checker (Scoreboard) Object (Continued)

```
        -- If there is a mismatch, diff will get the mismatches as
        -- a list of strings. Simulation will stop and the
        -- mismatches will be displayed.
        check that (diff.is_empty()) else
            dut_error("Packet not found on scoreboard", diff);

        -- If the match was successful, continue with the following
        -- actions
        out("Found received packet on scoreboard");
        -- On a match delete the matched packet
        exp_packets.delete(0);
    };

};


-- Extend the sbt_driver. When a packet is driven into the DUT
-- (indicated by the event packet_started) the
-- add_packet() method of the appropriate
-- scoreboard (indicated by the packets addr field) is called and the
-- cur_packet is copied to the exp_packets list
extend sbt_driver {
    -- Create a pointer to top level environment so we can reference
    -- the sbt_receivers
    parent_env: sbt_env; -- Pointer to the parent of sbt_driver
        keep parent_env == get_enclosing_unit(sbt_env);

    on packet_started {
    -- Add a copy of the cur_packet (rather than pointer) to
    -- the appropriate receiver instance.
    parent_env.sbt_receivers[cur_packet.addr].scoreboard.add_packet(cu
r_packet.copy());
    };
};
```

Example 13-4 *e* Code for the Data Checker (Scoreboard) Object (Continued)

```
-- Extend the sbt_receiver. Create an instance of the scoreboard.
-- When a packet was received (indicated by the event received_packet)
-- the scoreboard's check_packet() method is called which tries to
-- find a matching packet on the scoreboard's exp_packets list
extend sbt_receiver {
    scoreboard : sbt_scoreboard is instance; -- Instantiate scoreboard

    on pkt_received { -- At every pkt_received event
        scoreboard.check_packet(rcv_packet); -- Check packet against
                                             -- scoreboard element
    };
};
'>
```

13.5 Monitor Object

This section discusses the requirements and *e* code for the monitor object.

13.5.1 Requirements for the Monitor Object

The monitor object is implemented in the form of a protocol checker (*sbt_checker*). Requirements for the *sbt_checker* are as follows:

- If a packet is driven to the SBT with bad parity, expect that the output signal *err* is asserted 1 to 10 cycles after packet end.

- When the DUT signal, *suspend_data_in* rises, it must fall within 100 clock cycles. The *suspend_data_in* signal indicates that there is too much congestion in the router and it cannot accept more data.

- When the DUT signal, *vld_chan_X* (*vld_chan_0*, *vld_chan_1* or *vld_chan2*) rises, the input signal, *read_enb_X* (*read_enb_0*, *read_enb_1* or *read_enb_2*) should rise within 30 clock cycles.

- When the SBT DUT signal, *vld_chan_X* (*vld_chan_0*, *vld_chan_1* or *vld_chan2*) is asserted (high), and the *read_enb_X* (*read_enb_0*, *read_enb_1* or *read_enb_2*) is also asserted (high), the DUT signal *channelX* (*channel0*, *channel1* or *channel2*) should not be tri-stated (high-z).

Temporal expressions must be created to build the monitor object. A recommended technique is to create events for simple temporal expressions. These events are then combined to form complex temporal expressions.

13.5.2 *e* Code for the Monitor Object

Example 13-5 illustrates *e* code for the monitor object.

Example 13-5 *e* Code for the Monitor Object

```
File contains e code for the monitor object.
In this file the monitor object is not a separate
unit. In this case it simply an extension of the sbt_driver
and sbt_receiver objects.
<'
-- Extend the sbt_driver for protocol checks
extend sbt_driver {

-- Declare events bad_parity and err_rise
event bad_parity is true(cur_packet.packet_kind == BAD)@packet_ended;
event err_rise is rise('err')@clk_fall;

-- "If a packet is driven to the SBT with bad parity, expect that the
-- output signal err is asserted 1 to 10 cycles after packet end"
expect @bad_parity => {[0..9]; @err_rise} @clk_fall else
        dut_error("A packet with bad parity was driven into the SBT ",
        "but the err signal was not asserted.");

-- Declare simple events
event suspend_data_rise is rise('suspend_data_in')@clk_fall;
event suspend_data_fall is fall('suspend_data_in')@clk_fall;
-- "When the SBT DUT signal, suspend_data_in rises, it must fall
-- within 100 clock cycles."
expect @suspend_data_rise => {
        [0..99];
        @suspend_data_fall;
    } @clk_fall else
        dut_error("The suspend_data_in signal was asserted for more ",
        "than 100 clock cycles.");
};

-- Extend the sbt_receiver for protocol checks
extend sbt_receiver {

-- Use computed names using the port_no field since we have multiple
-- receivers
event vld_chan_rise is rise('vld_chan_(port_no)')@clk_fall;
event read_enb_rise is rise('read_enb_(port_no)')@clk_fall;
```

Example 13-5 *e* Code for the Monitor Object (Continued)

```
-- "When the SBT DUT signal, vld_chan_(port_no) rises, the input
-- signal read_enb_(port_no) should rise within 30 clock cycles."
expect @vld_chan_rise => {
      [0..29];
      @read_enb_rise;
   } @clk_fall else
      dut_error("For port ", port_no, " the read_enb_", port_no,
      " signal did not rise within 30 cycles of the rise of
      vld_chan_", port_no);

-- Event when both vld_chan_X and read_enb_X are asserted
event vld_chan_read_enb_assr is true(('vld_chan_(port_no)' == 1'b1)
      and ('read_enb_(port_no)' == 1'b1))@clk_fall;

-- Use @z to check if any bits in data signal are high-z
event data_high_z is true('data@z' != 8'h00)@clk_fall;

-- "When the SBT DUT signal, vld_chan_(port_no) is asserted (high),
-- and the "read_enb_(port_no)" is also asserted (high), the output
-- DUT channel(port_no) should not be tri-stated (high-z)."
expect not(@data_high_z) @vld_chan_read_enb_assr else
      dut_error("For port ", port_no, " the data had a tri-state ",
      "value during the time that vld_chan_", port_no,
      " was asserted.");
};
'>
```

13.6   Coverage Object

This section discusses the requirements and *e* code for the coverage object. Coverage is implemented in two forms:

- By extending the packet object to add coverage on the items of the packet struct.
- By creating a separate coverage object (*sbt_dut_cover*) to perform coverage on the state machine.

13.6.1   Requirements for Adding Coverage to the Packet Object

Requirements for the packet object are as follows:

- Add an event *new_packet* to use for coverage of packet items.
- Extend the *sbt_packet* struct and cover the fields *addr*, *len*, *packet_kind*, and *payload_size*.

- For the *len* field, use the **ranges{}** coverage item option to create three buckets of equal range size.

13.6.2 *e* Code for Adding Coverage to the Packet Object

Example 13-6 shows *e* code for adding coverage to the jacket object.

Example 13-6 *e* Code for Adding Coverage to the Packet Object

```
<'
import sbt_packet;

extend sbt_packet {
    -- Extend sbt_packet with event for coverage
    event new_packet;
    -- Coverage will happen on the event new_packet
    cover new_packet is { -- Coverage on fields of packet
        -- Use ranges for the len field
        item len using ranges = {
            range([1..20], "SMALL");
            range([21..40], "MEDIUM");
            range([41..63], "LARGE");
        };
        item addr; -- Cover item addr
        item packet_kind; -- Cover item packet_kind
        item payload_size; -- Cover item payload_size
    };
};

-- Extended the sbt_driver to emit the new_packet coverage event for
-- the current packet at the time it is driven into the DUT
extend sbt_driver {
    on packet_started {
        emit cur_packet.new_packet;
    };
};

'>
```

13.6.3 Requirements for the Coverage Object

The coverage object is implemented in the form of a scoreboard (*sbt_dut_cover*). The *sbt_dut_cover* object performs coverage on the DUT state machine. Requirements for the *sbt_dut_cover* object are as follows:

- Define an enumerated type for the states of the DUT finite state macine (FSM).

- Add a coverage item of type *fsm_state* to cover the HDL DUT state.

- Add a transition coverage item for the state register. Define the illegal transitions in the DUT state machine. (Here it is easier to specify illegal transitions as a **not** of legal transitions.)

13.6.4 *e* Code for the Coverage Object

Example 13-7 presents *e* code for the coverage object.

Example 13-7 *e* Code for the Coverage Object

```
File contains e code for the coverage object.
Record the DUT state machine coverage
- Create a new enum type fsm_state for the SM (has to match
- the HDL definition)
- Create a new coverage group for the SM coverage
- Cover the DUT state vector
- Create a transition coverage item, define all the illegal
- transitions
  (Define all the legal transitions and invert it using
   the "not" operator)
<'

-- Define enumerated type for fsm_state
-- Order has to be the same as in the HDL source
type fsm_state : [ADDR_WAIT, DATA_LOAD, PARITY_LOAD,
                  HOLD_STATE, BUSY_STATE];

unit sbt_dut_cover {
 -- Event sbt_sm happens every time clock_rise happens
    event sbt_sm is rise('clock')@sim;

    cover sbt_sm using text = "SBT State Machine" is {

        -- Assign the DUT state vector to the item "state"
        item state : fsm_state = 'router1/in_port/state_r';

        -- Legal/illegal states have to be complete!!
        -- We use all the legal transitions and invert them
        transition state using illegal = not ( -- of legal transitions
            (prev_state == ADDR_WAIT and
                ((state == ADDR_WAIT) or (state == DATA_LOAD) or
                (state == BUSY_STATE))) or
            (prev_state == BUSY_STATE and
                ((state == BUSY_STATE) or (state == DATA_LOAD))) or
            (prev_state == DATA_LOAD and
                ((state == DATA_LOAD) or (state == HOLD_STATE) or
                (state == PARITY_LOAD))) or
            (prev_state == HOLD_STATE and
                ((state == HOLD_STATE) or (state == DATA_LOAD) or
                (state == PARITY_LOAD))) or
            (prev_state == PARITY_LOAD and state == ADDR_WAIT)
```

Example 13-7 *e* Code for the Coverage Object (Continued)

```
       ); -- end transition

    }; -- end coverage

}; -- end unit

-- This configuration will enable coverage (Default is disabled)
extend sys {
    setup() is also {
        set_config(cover, mode, on); -- Other coverage options
                                     -- are available
    };
};

'>
```

13.7 Environment Object

This section discusses the requirements and *e* code for the environment object. The environment object is introduced to make the verification system of the router more transportable to other environments. The environment object is then instantiated under **sys**. Otherwise, the components under the environment could be instantiated directly under the **sys** struct.

13.7.1 Requirements for Environment Object

The primary requirement for the environment object (*sbt_env* **unit**) is to build the hierarchy as shown in Figure 13-1. Note that *sbt_env* unit is instantiated under the predefined **sys** struct.

Figure 13-1 Environment Object Hierarchy

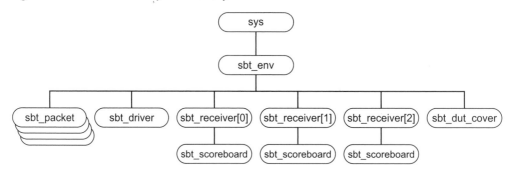

13.7.2   *e* Code for the Environment Object

Example Figure 13-8 displays *e* code for the environment object.

Example 13-8 *e* Code for the Environment Object

```
File contains e code for the environment object.
<'
-- Import all files
import sbt_packet;
import sbt_driver;
import sbt_scoreboard;
import sbt_receiver;
import sbt_checker;
import sbt_packet_cover;
import sbt_dut_cover;

struct sbt_env {
    -- Instantiate one sbt_driver
    sbt_driver : sbt_driver is instance;
    keep sbt_driver.hdl_path() == "~/top";

    -- Create a list of sbt_receivers (3 are needed)
    -- One per each channel. Each receiver in turn
    -- also instantiates the sbt_scoreboard
    sbt_receivers[3] : list of sbt_receiver is instance;
    -- Assign the numbers 0, 1 and 2 to the number field of each
    -- receiver
    keep for each in sbt_receivers {
        .hdl_path() == "~/top"; -- define hdl_path() of each
                                -- receiver
        .port_no == index; -- set port_no of each receiver
                           -- 0, 1, 2.
    };
```

Example 13-8 *e* Code for the Environment Object (Continued)

```
    -- Create one instance of sbt_dut_cover
    sbt_dut_cover : sbt_dut_cover is instance;
    keep sbt_dut_cover.hdl_path() == "~/top"; -- Set hdl_path
};

-- Create an instance of sbt_env object in
-- the top-level sys struct.
extend sys {
    sbt_env : sbt_env is instance;
};


'>
```

13.8 Test Scenarios

This section discusses the requirements and the *e* code required to implement the tests in the "Test Plan" on page 241. Each test is simply a small file containing *e* code. To run the test, compile the test file with the rest of the *e* files.

13.8.1 Requirements for Test Scenario 1 (Distribution of Packets)

In this scenario, create packets with a certain probability distribution. There should be a mix of packets of different lengths. There should also be a mix of good and bad parity packets. To do so, extend the **struct** *sbt_packet* as follows:

- The packet lengths should be distributed with 10% SMALL, 20% MEDIUM, and 70% LARGE.

- The packet kind must be 70% with good parity and 30% with bad parity.

Extend the **struct sys** as follows:

- Create exactly 50 packets and inject them into the DUT.

13.8.2   *e* Code for Test Scenario 1 (Distribution of Packets)

Example 13-9 presents *e* code for the test scenario 1 (distribution of packets).

Example 13-9 *e* Code for Test Scenario 1 (Distribution of Packets)

```
File contains test that distributes the types
of packets in a probabilistic distribution.
10% SMALL, 20% MEDIUM, 70% LARGE
70% Good parity, 30% Bad parity
<'
import sbt_env; -- Import all units instantiated in the environment.

extend sbt_packet {
    keep soft payload_size == select {   -- Weighted constraint for the
                                         -- payload_size field
        10: SMALL; -- Probability 10%
        20: MEDIUM; -- Probability 20%
        70: LARGE; -- Probability 70%
    };

    keep soft packet_kind == select {-- Weighted constraint for the
                                     -- packet_kind field
        70: GOOD; -- Probability 70%
        30: BAD; -- Probability 30%
    };
};

extend sys {
    keep sbt_env.sbt_driver.no_of_pkts == 50; -- Set number of packets
    -- This field will decide how many packets get generated in the
    -- sbt_driver and are injected into the DUT.
};
'>
```

13.8.3   Requirements for Test Scenario 2 (Protocol Violation)

In this scenario, create a test that will make the *sbt_receiver* assert the *read_enb_X* (*read_enb_0*, *read_enb_1* or *read_enb_2*) more than 30 *clock* cycles after *vld_chan_X* (*vld_chan_0*, *vld_chan_1* or *vld_chan2*) is asserted. This will violate the protocol. To do so, extend the **struct** *sbt_receiver* as follows:

- Set the *rcv_delay* field to be 40. This will violate the protocol.

Note the simplicity of *e* tests. The test files are simply small files with constraints.

13.8.4 *e* Code for Test Scenario 2 (Protocol Violation)

e code for test scenario 2 (protocol violation) is show in Example 13-10.

Example 13-10 *e* Code for Test Scenario 2 (Protocol Violation)

```
File contains test that violates the protocol.
This is done by setting a simple constraint
that will violate the protocol for the latency
between vld_chan_X and read_enb_X.
<'
extend sbt_receiver {
    keep rcv_delay == 40;    -- Violate the DUT output protocol rule
                             -- Latency is 40, violates the maximum
                             -- limit of 30
};
'>
```

13.9 Summary

- The *e* source code for all components of the router verification environment was discussed.

- The verification environment consists of data object (**struct** *packet*), driver object (**unit** *sbt_driver*), receiver object (**unit** *sbt_receiver*), data checker object (**unit** *sbt_scoreboard*), monitor object (extend of **unit** *sbt_driver* and **unit** *sbt_receiver*), coverage object (**unit** *sbt_dut_cover*), and environment object (**unit** *sbt_env*).

- A typical test plan covers all possible test scenarios for a DUT. In this chapter, we have only discussed two simple test scenarios. Tests in *e* are very simple constraint files.

- Using the concepts in this chapter, a reader can build more complex verification environments.

PART 4

Advanced Verification Techniques with *e*

14 **Coverage Driven Functional Verification**
Traditional Verification Methodology, Why Coverage?,
Coverage Approaches, Functional Coverage Setup,
Coverage Driven Functional Verification Methodology

15 **Reusable Verification Components (*e*VCs)**
About *e*VCs, *e*VC Architecture, *e*VC Example

16 **Interfacing with C**
C Interface Features, Integrating C Files, Accessing the
e Environment from C, Calling C Routines from *e*,
Calling *e* Methods from C, C Export, Guidelines for
Using the C Interface, Linking with C++ Code

CHAPTER 14

Coverage-Driven Functional Verification

Traditional verification techniques did not include coverage measurement as a key component until the later stages of the project cycle. Coverage measurement is applied only after the testbench and the design are mature. Coverage measurement is used as a checklist item at the end of the project. With e, coverage measurement is now an integral part of the verification methodology. This integration enables new methodologies that can dramatically increase coverage quality and greatly decrease the number of simulations required to achieve such coverage. This chapter focuses on the coverage-drivencoverage-driven functional verification methodology.

Chapter Objectives

- Describe a traditional verification methodology, i.e., test-driven verification.
- Define the problems with a traditional verification methodology.
- Explain coverage-driven functional verification.
- Understand the advantages of coverage-driven functional verification.
- Describe complementary coverage approaches.

14.1 Traditional Verification Methodology

In a traditional verification methodology, the verification engineer goes through the following steps to verify a DUT (block, chip, or system):

1. Create a test plan that contains directed tests for the DUT based on the engineer's knowledge of the design specification.

2. Write the directed tests. The engineer typically spends a lot of manual effort in writing these directed tests. Since the DUT is still evolving at this point, it is impossible to pre-

dict where the bugs might be. It is possible that a certain directed test may not uncover any bugs. Moreover, many directed tests may overlap with other directed tests, thus testing the same functionality.

3. Run the directed tests to find bugs in the DUT. Since the directed tests verify specific scenarios, only bugs pertaining to those specific scenarios are detected. Other scenarios are left uncovered.

4. Add more directed tests if necessary to cover new scenarios. Engineer spends more manual effort thinking about new scenarios that need to be tested.

5. Run these additional directed tests to find more bugs in the DUT. Steps 4 and 5 are run until the engineer is convinced that enough directed testing has been done. However, the measurement of adequacy is still very ad hoc.

6. Random testing is initiated with some form of a random stimulus generator after multiple iterations of steps 4 and 5 are performed.

7. Random testing uncovers bugs that were not detected originally by directed tests. Random testing often catches corner cases that were missed by the verification engineer. The bugs that are uncovered are therefore fixed in a very late stage of the verification process.

8. Functional coverage is initiated after multiple iterations of steps 6 and 7 are performed. Functional coverage is run mainly in the post-processing mode and it provides results on the values of interesting items and state transitions. Random simulations are run until desired functional coverage levels are achieved.

9. After steps 1 through 8 are performed and the coverage results are satisfactory, the verification is considered complete.

Figure 14-1 shows the traditional verification process outlined above.

Figure 14-1 Traditional Verification Methodology

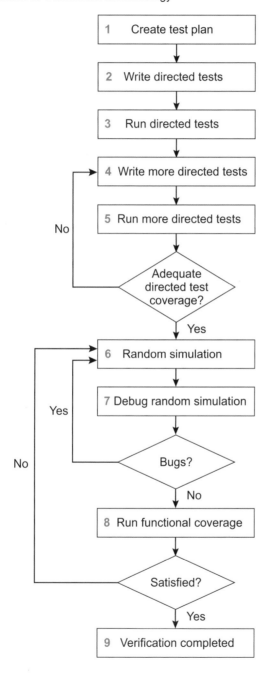

14.1.1 Problems with a Traditional Verification Methodology

The traditional verification methodology shown in Figure 14-1 has key productivity and quality issues:

- The verification effort starts with directed testing. Directed tests require intense manual effort, are time consuming to create, and are extremely difficult to maintain. Thus, rather than focus on what areas to verify, the engineer spends most of the time figuring out how to write directed tests.

- At this stage the DUT is rapidly changing in both functionality and implementation. Therefore, the verification engineer might focus on writing directed tests for areas that are not prone to bugs and hence waste valuable verification time.

- There is no quantifiable metric about how much directed testing is enough. Currently, verification engineers stop when they have written a large number of tests and they think that they have adequate coverage or they run out of time. This process is very unstructured and ad hoc. This reduces the quality of the DUT verification.

- Random simulation uncovers bugs that are not caught by directed tests. However, if directed tests were not run, random simulation would have also caught most of the bugs that directed tests caught. Therefore, the verification engineer spent a lot of manual effort finding bugs that would have been caught very easily with random simulation.

- At some point, the engineer is satisfied that there are no more bugs from random simulation. Often, this decision is based on the fact that random simulation ran for a certain number of days without a bug. Although this means that the DUT is reasonably stable, it does not guarantee that all interesting corner cases have been covered.

- It is very hard to control random simulation. A lot of time is spent tuning the random simulator to catch corner cases.

- Coverage is run towards the end of the verification cycle. It is mainly intended as a checklist item to ensure that decent coverage has been achieved. However, it is not used as a feedback mechanism to tune the quality of tests.

Thus, it is clear that many aspects of the traditional verification methodology are inefficient and very manual in nature. The problems with this approach lead to lots of wasted verification time and suboptimal verification quality.

14.1.2 Reasons for using Traditional Verification Methodology

Although there are obvious problems with the traditional verification methodology, verification projects have persisted with this approach over the years. There are some important reasons for this dependence on the traditional verification methodology:

- Commercial random simulation tools were not easily available. Moreover, these tools were not easily customizable.

- Random simulation tools were often developed in-house by companies. But these tools required a massive maintenance effort and therefore were affordable only to big companies.

- Random simulation tools were not integrated into the verification environment. Therefore, it took a lot of time and effort for verification engineers to integrate these tools into their environment.

- Random simulation tools did not afford a high degree of control. Directed-random (controlled random) simulation tools were not available.

- Coverage tools were not integrated into the environment. A verification engineer needed to spend a lot of time and effort integrating the coverage tool.

- Coverage tools only provided post-processed reports. The process of taking coverage reports and using that feedback to generate new tests was not smooth.

Due to these problems, verification engineers could never take advantage of the coverage-driven functional verification approach that can provide more productivity and higher verification quality. Most of their time was spent on integrating the tools and working with inefficient verification flows, rather than on focusing on DUT verification.

14.2 Why Coverage?

Coverage information is highly valuable because it ensures that verification is performed in the most effective manner. Coverage measurement provides the following information:

- Coverage provides an important feedback loop for improving the quality of stimulus and for rigorous checking.

- Coverage also provides a feedback loop for improving the performance of the DUT.

- Coverage also helps the engineer simulate efficiently by optimizing the regression suite and by running only what is needed.

The following sub-sections discuss how coverage can improve various aspects of the verification environment.

14.2.1 Coverage Improves Stimulus

The quality of the stimulus can be improved by measurement of coverage of the input fields. If a particular input combination is missing, coverage can immediately report the missing combination. Such feedback is important to improve the quality of the stimulus. Figure 14-2 shows how coverage can help improve stimulus quality.

Figure 14-2 Coverage Improves Stimulus

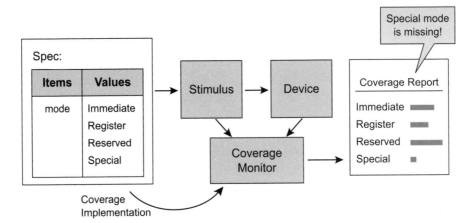

14.2.2 Coverage Improves Checking

The quality of checking can also be improved by measurement of coverage. Often, if the monitor does not report an error, it is assumed that the DUT is working properly. However, it is possible that the DUT did not enter a state where it would encounter a bug. For example, Figure 14-3 shows that if the DUT never enters the *!ready* state, the checker will never report an error. Since coverage reports that the DUT never entered the *!ready* state, the verification engineer knows that the checking is not complete.

Figure 14-3 Coverage Improves Checking

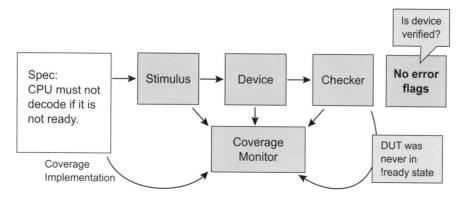

14.2.3 Coverage Verifies the DUT

Coverage also verifies that the DUT works for a varied distribution of results. Thus, the DUT is not verified for a single case but a wide distribution of cases. This helps better verification of the DUT in different scenarios that the DUT will face in real-life environments. For example, Figure 14-4 shows how a DUT can be tested for arbitration fairness (reasonable latencies to get grants) and a reasonable latency distribution.

Figure 14-4 Coverage Verifies the DUT

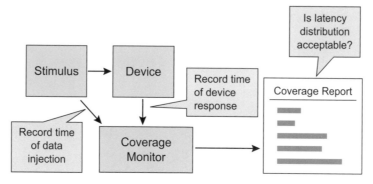

14.2.4 Coverage Measures Progress

Coverage ensures that every additional test that is run improves the coverage. This guarantees that each test is effective towards verifying an additional part of the DUT. This guarantees continuous progress. Figure 14-5 shows how the verification engineer can focus attention on scenario d, if scenarios $a, b,$ and c are already covered.

Figure 14-5 Coverage Measures Progress

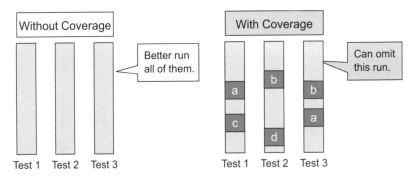

14.2.5   Coverage Enables Regression Optimization

Coverage allows the optimization of a regression run. Tests that repeat scenarios already covered in previous runs can be omitted. Let us assume that a regression suite contains only three tests. Figure 14-6 shows that without coverage all three tests have to be run because there is no insight into which scenarios are covered by the tests. However, when these tests are run with coverage, we realize that test 3 repeats scenarios a and b that are already covered by tests 1 and 2. Hence, test 3 should be omitted from the regression because it does not test any new scenario but only wastes valuable computing resources.

Figure 14-6 Coverage Enables Regression Optimization

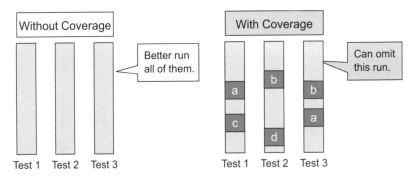

14.3   Coverage Approaches

There are many coverage approaches. This section discusses the popular coverage approaches.

14.3.1   Toggle Coverage

One of the oldest coverage measurements, toggle coverage, was historically used for manufacturing tests. Toggle coverage monitors the bits of logic that have toggled during simulation. If a bit doesn't toggle from 0 to 1, or from 1 to 0, it hasn't been adequately verified. Toggle coverage does not ensure completeness. Moreover, toggle coverage is very low level and it may be cumbersome to relate a specific bit to a test plan item. However, toggle coverage is often used to provide additional confidence that the design has been exercised thoroughly.

14.3.2 Code Coverage

A popular coverage metric is code coverage. The basic assumption of code coverage is that unexercised code potentially bears bugs, i.e., the code is guilty until proven innocent. Code coverage checks how well your Register Transfer Level (RTL) code was exercised. Figure 14-7 shows how code coverage measures the level of activity of the RTL code.

Figure 14-7 Code Coverage

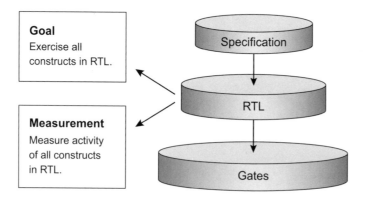

14.3.3 Functional Coverage

Functional coverage measures the activity of the DUT against goals derived from the specification. Functional coverage measures how completely the functionality of the DUT was verified. Figure 14-8 shows how functional coverage measures activity as specified coverage points in the DUT.

Figure 14-8 Functional Coverage

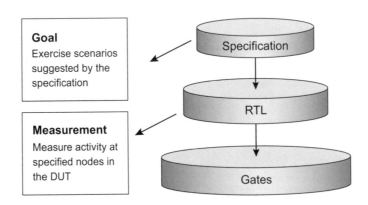

14.3.4　Functional Coverage vs. Code Coverage

It is tempting to use only code coverage or functional coverage. However, both coverage techniques are necessary to ensure adequate coverage of the DUT. Figure 14-9 shows the comparison between functional coverage and code coverage. In this book, we focus on functional coverage.

Figure 14-9 Functional vs. Code Coverage

	Code Coverage	Functional Coverage
Goals	Derived from implementation	Derived from the specification
Effort to implement	Automatic	Well-defined process for creation of metrics
Misses important goals if...	Implementation omits desired functionality	Process omits important metrics
Susceptible to false optimism	Due to "occurred but not verified" problem	Careful specification avoids "occurred but not verified" problem

14.4 Functional Coverage Setup

Functional coverage has to be planned thoroughly. This section discusses the overall methodology for setting up coverage.

14.4.1 Coverage Process

One beings the coverage process by deriving information from the specification regarding items that need to be covered. Based on that information, a coverage model is created. The coverage model is run in conjunction with the test base and it provides coverage reports. These reports are analyzed to do one of the following:

- Refine the test base.
- Refine the *e* code.
- Refine the coverage model, i.e., the items that need to be covered.

Figure 14-10 shows a typical coverage process.

Figure 14-10 Coverage Process

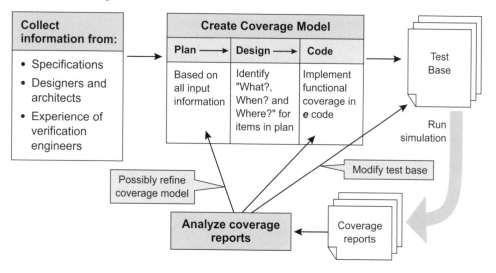

14.4.2 Coverage Plan

A coverage plan is a specification of a coverage model. A coverage plan contains information to be collected about attributes of the DUT, interactions between different attributes of the DUT, and a sequence of occurrences. A coverage plan also contains information about when these items should be collected. Figure 14-11 shows the example of a typical coverage plan.

Figure 14-11 Coverage Plan

Attribute	Important values	When valid
mode	Immediate, register, reserved, special	Upon fetch
Asb-to-drdy delay	Zero, short (1–5 cycles), long (over 5 cycles)	Upon drdy

Interaction	Important values	When valid
mode, opcode	[Immediate, register]	Upon fetch

Sequence	Important values	When valid
Last 4 opcodes	All combinations of opcode categories	Upon fetch

14.4.3 Coverage-Driven Verification Environment

Once the coverage process and the coverage plan are set up, the coverage-driven verification environment needs to be planned carefully. A good coverage environment includes automatic stimulus generation, self-checking, and a coverage monitor as an integral part of the environment. Optional manual guidance can be provided to direct the environment towards certain interesting areas. Figure 14-12 shows the setup of a typical coverage-driven verification environment.

Figure 14-12 Coverage-Driven Verification Environment

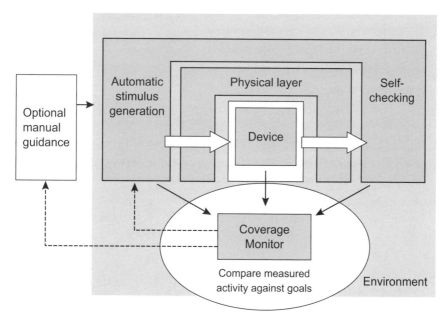

14.5 Coverage Driven Functional Verification Methodology

A coverage-driven functional verification approach combines directed-random simulation with coverage to solve the problems with a traditional verification methodology. This approach greatly enhances productivity of the engineers and the quality of the DUT.

14.5.1 What Does *e* Provide?

Both directed-random generation as well as coverage are provided in a unified manner in *e* syntax.

- **keep** constraints control directed-random generation.
- The **cover** struct members provide coverage functionality.

These *e* constructs enable a new verification paradigm that was not possible before with older tools.

14.5.2 The New Verification Paradigm

In a coverage-driven functional verification methodology, the verification engineer goes through the following steps to verify a DUT (block, chip, or system):

1. Create a test plan that identifies all possible behavior of the DUT to be verified. Define coverage requirements on interesting items and state transitions in the DUT. This meth-

odology requires that the test plan and the coverage requirements be very thorough because it sets up a reference for the measurement of 100% coverage.

2. Set up the verification environment with constraints (**keep** and **keeping** constraints) from the DUT specification and test plan. Specify coverage on interesting items and state transitions in the DUT using the **cover** struct member.

3. Run many directed-random simulations with different seed values. The constraints will ensure that the randomness is controlled and all interesting areas are touched. Most obvious bugs will be uncovered by random simulation with minimal manual effort.

4. Run coverage with random simulations to find out how many DUT areas have been tested. Coverage will tell what DUT cases have not been reached. Run random simulations with different seeds until no more DUT areas can be reached.

5. Refine the verification constraints of the verification environment to verify device behavior not previously possible. Directed tests are simply tests with lots of constraints. There is some manual effort but it is only focused on areas that are not reachable by random simulation.

6. Rerun functional coverage measurement with directed tests to ensure that all corner cases were covered.

7. Repeat steps 5 and 6 until you reach 100% coverage.

8. When you achieve 100% coverage, the verification is considered complete.

Figure 14-13 shows the coverage-driven functional verification methodology outlined in this section. This new verification paradigm is enabled because both directed-random simulation as well as coverage can be used as a combination from the very beginning of the verification effort. This was not possible earlier due to lack of integrated tools to do both directed-random simulation and functional coverage.

Figure 14-13 Coverage Driven Verification Methodology

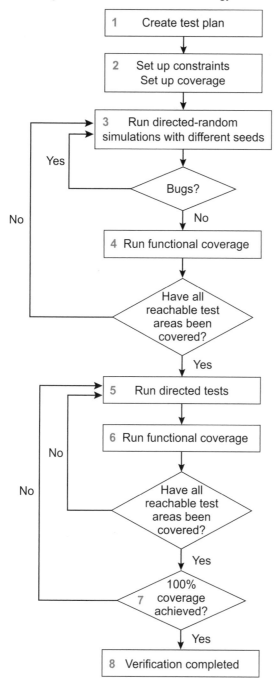

14.5.3 Advantages of Coverage Driven Functional Verification

Using the coverage-driven functional verification paradigm has numerous productivity and quality advantages.

- Minimal time is spent on directed tests at the beginning of the verification cycle. Usually, directed tests consume about 70-80% of the verification life cycle. That time is significantly shortened because directed-random simulations can be fired with minimal effort and can be completed in a short time.

- Running directed-random simulations with multiple seeds at the beginning of the verification cycle reaches 80-90% of the interesting test cases. There is no need to spend the manual effort in creating directed tests to reach these test cases.

- Directed tests are created only for the 10-20% hard-to-reach test cases that cannot be exercised by directed-random simulation. However, in this case, the goal of each directed test is very specific.

- The number of directed tests is small. Therefore, maintenance of the verification environment is much simpler.

- Since functional coverage is involved at each step, there is clear feedback about the effectiveness of each test, i.e., whether the test gives additional coverage. If the test increases coverage, it is retained, otherwise, it is discarded. This maximizes the engineer's productivity in creation and maintenance of tests.

- The ultimate goal of any verification environment is to create a bug-free DUT with 100% coverage. With coverage-driven functional verification, there is a clear metric that tells when the verification is complete. This maximizes the quality of the DUT.

14.6 Summary

- In a traditional verification methodology, directed tests are written before directed-random simulations. Coverage is usually performed as a final step to ensure completeness of design verification.

- Coverage information is highly valuable because it ensures that verification is performed in the most effective manner. Coverage provides an important feedback loop for improving the quality of stimulus, for rigorous checking, for improving the performance of the DUT, and for optimal regression testing.

- Both functional coverage and code coverage are required to ensure adequate coverage for the verification of the DUT.

- Setup of functional coverage requires a well-designed coverage process, a coverage plan, and a coverage-driven environment that includes automatic stimulus generation, self-checking, and coverage monitoring.

- In a coverage-driven functional verification methodology, directed-random simulations are run along with the functional coverage tool to test the easy-to-reach test cases. Directed tests are written only for the hard-to-reach test cases. Coverage is measured at each step in this approach.

- In a coverage-driven functional verification methodology, the final goal is clear and measurable, i.e., a bug-free DUT with 100% coverage. Therefore, this methodology requires that the test plan and the coverage requirements are very thorough because it sets up a reference for the measurement of 100% coverage.

CHAPTER 15

Reusable Verification Components (*e*VCs)

One of the key factors for reusing code is arranging it as an independent and easy-to-use code package. When verification code is developed in *e*, the reusable package is typically organized as an *e*VC (*e* Verification Component). The methodology for developing these *e*VCs is known as the *e* Reuse Methodology (*e*RM). Following the *e*RM ensures that the *e* code developed can be used seamlessly across block, chip, and system level environments and also across projects. This chapter introduces the reader to the *e*RM. Details of the *e*RM are left to the *e*RM *Reference Manual*.

Chapter Objectives

- Describe *e*VCs.
- Define the *e* Reuse Methodology (*e*RM).
- Describe the architecture of a typical *e*VC.
- Understand a complete example of an *e*VC.

15.1 About *e*VCs

This section provides an introduction to *e*VCs.

15.1.1 What Are *e*VCs?

An *e*VC™ is an *e* Verification Component. It is a ready-to-use, configurable verification environment, typically focusing on a specific protocol or architecture (such as Ethernet, AHB, PCI, or USB).

Each *e*VC consists of a complete set of elements for stimulating, checking, and measuring coverage information for a specific protocol or architecture. You can apply the *e*VC to your device under test (DUT) to verify your implementation of the *e*VC protocol or architecture. *e*VCs expedite creation of a more efficient testbench for your DUT. They can work with both Verilog and VHDL devices and with all HDL simulators that are supported by Specman Elite. *e*VCs can work with any simulator that supports the *e* language.

You can use an *e*VC as a full verification environment or add it to a larger environment. The *e*VC interface is viewable and hence can be the basis for user extensions. It is recommended that such extensions be done in a separate file. Maintaining the *e*VC in its original form facilitates possible upgrades.

An *e*VC implementation is often partially encrypted, especially in commercial *e*VCs where authors want to protect their intellectual property. Most commercial *e*VCs require a specific feature license to enable them.

Following is a partial list of possible kinds of *e*VCs:

- *Bus-based *e*VCs* (such as PCI and AHB).
- *Data-communication *e*VCs* (for example, Ethernet, MAC, Datalink).
- *CPU/DSP *e*VCs*.
- *Higher-level protocol *e*VCs* (TCP/IP, HTTP). These usually sit on top of other *e*VCs.
- *Platform *e*VCs* (that is, an *e*VC for a specific, reusable System-on-Chip (SoC) platform, into which you plug *e*VCs of various cores).
- *Compliance test-suite *e*VCs*. These are tests (and perhaps coverage definitions and more) that demonstrate compliance to a protocol. For example, there could be a PCI compliance *e*VC in addition to the basic PCI *e*VC.
- *HW/SW co-verification *e*VCs*, such as an *e*VC dedicated to verifying a HW/SW environment using a particular RTOS/CPU combination.

15.1.2 *e*VCs as Plug and Play Components

Ideally, *e*VCs must be plug and play components in the sense that a new verification environment can be constructed from *e*VCs that were not initially planned to work together. It should also be possible to do this hierarchically.

Following are the benefits of making *e*VCs plug and play:

- Promotes code sharing
- Offers customers a convenient, ready-to-use product
- Makes *e*VC creation easier and faster

15.1.3 *e*VC Reuse Requirements

Table 15-1 lists the requirements that are essential for *e*VC reuse. The *e*RM specifies in detail how to implement each requirement.

Table 15-1 *e*VC Reuse Requirements

No interference between *e*VCs	• No name space collision • No complex search path or directory dependencies • Handling dependencies on common modules • No dependencies on different versions of Specman Elite and its utilities • No timing dependencies • No dependencies on global settings
Common look and feel, similar activation, similar documentation	• Common way to install *e*VCs • Common way to patch *e*VCs • Common tracing and debugging • Handling DUT errors • Getting *e*VC identification • Waveform viewer data • Custom visualization • Common way of specifying simulator-specific material • Common way to do backdoor initialization • Common programming interface to standard blocks • Common *e*VC taxonomy • Common style of documentation
Support for combining *e*VCs (control, checking, layering, and so on)	• Common way to configure *e*VCs • Common way to write tests • Common way to create sequences • Common way to do checking • Combined determination of end of simulation • Common way to do layering of protocols • Common way to do combined coverage
Support for modular debugging	• Understanding combined constraints • Reconstructing the behavior of a single *e*VC in the verification environment
Commonality in implementation	• Common data structures • Common *e*VC simulation testing methodology • Common way to use ports and packages

15.2 *e*VC Architecture

Each *e*VC relates to different protocols, architectures, and designs. Therefore, one *e*VC can vary a lot from another *e*VC. Still, there is a lot of commonality in the design of various *e*VCs. In this section, we will investigate the commonality and some of the main differences between typical *e*VCs.

In this section, we take a sample *XSerial* *e*VC to show how a typical *e*VC should be constructed. The *XSerial* *e*VC is an example of how to code a general-purpose *e*VC for a point-to-point protocol. The chosen protocol is deliberately simple so that attention is focused on the *e*VC methodology rather than the difficulties in coding complex BFMs. The protocol is a synchronous, full-duplex serial protocol with a frame consisting of 8-bit data, 2-bit address, and 2-bit frame-kind. This chapter does not focus on the details of the protocol, but instead elaborates the structure of the *XSerial* *e*VC.

15.2.1 DUT and *e*VC

Figure 15-1 below shows a sample DUT that we will use to demonstrate some *e*VCs. This DUT has two external interfaces: a bus interface, and a serial interface. Each of these interfaces can interact with the external world, and therefore we attach an *e*VC to exercise and interact with each of them.

Figure 15-1 Example DUT

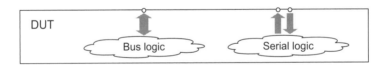

We will start by looking at the serial interface. This interface is composed of a receive port and a transmit port. The *e*VC attached to this interface is the *Xserial* *e*VC. A dual-agent implementation for the *XSerial* *e*VC is shown in Figure 15-2.

Figure 15-2 XSerial *e*VC—Dual-Agent Implementation

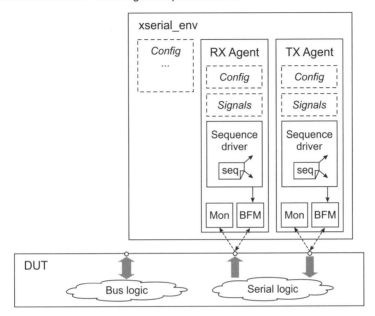

The XSerial *e*VC is encapsulated in the rectangle marked *xserial_env*. For each port of the interface, the *e*VC implements an agent. These agents can emulate the behavior of a legal device, and they have standard construction and functionality. Each env also has a group of fields, marked in Figure 15-2 as *Config*. This allows configuration of the env's attributes and behavior. The agents are units instantiated within the env.

In Figure 15-2, notice that the BFMs have bidirectional arrows to the DUT. This signifies the fact that they can both drive and sample DUT signals. Monitors have unidirectional arrows pointing from the DUT to them. This signifies that they can only sample data from the DUT. Monitors cannot drive input into the DUT.

In this representation of the *e*VC, there are two types of agents as shown in Table 15-2.

Table 15-2 Types of Agents in an *e*VC

RX agent	A receive agent that can collect data from the DUT's transmit port
TX agent	A transmit agent that can send data to the DUT's receive port

These agents are constructed in a standard way. The components of an *e*VC agent are described in Table 15-3.

Table 15-3 Components of an *e*VC Agent

Config	A group of fields that allow configuration of the agent's attributes and behavior.
Signals	A group of unit members that represent the hardware (HW) signals that the agent must access as it interacts with the DUT. Currently, the signals are implemented as string fields, all prefixed with "**sig_**".
Sequence Driver	This is a unit instance that serves as a coordinator for running user-defined test scenarios (implemented as sequences).
BFM	Bus Functional Model—a unit instance that interacts with the DUT and drives or samples the DUT signals.
Monitor	A unit instance that passively monitors (looks at) the DUT signals and supplies interpretation of the monitored activity to the other components of the agent. Monitors can emit events when they notice interesting things happening in the DUT or on the DUT interface. They can also check for correct behavior or collect coverage.

The *XSerial e*VC could also be implemented using a single-agent architecture as shown in Figure 15-3 below.

Figure 15-3 XSerial *e*VC—Single-Agent Implementation

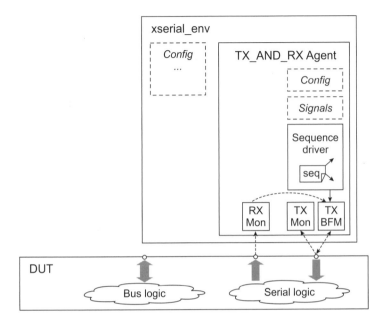

Both the dual-agent and the single-agent architectures are good implementations. Depending on the specifics of the protocol with which the *e*VC deals, you might prefer one over the other.

In the case of the single-agent *XSerial* protocol, the Receive (RX) direction is completely passive and involves no driving of signals. Thus, there is no need to have a BFM and a sequence driver in the RX agent. However, in the single-agent *XSerial* protocol, the Transfer (TX) agent behavior also depends on flow control frames received by the RX agent. This means that the RX agent must communicate frequently with the TX agent to tell it when it has received such frames.

In the single-agent *XSerial e*VC, the RX agent is significantly simpler than the TX agent. Therefore, it is better to implement a single-agent *e*VC to efficiently model the flow control mechanism, given that the flow control involves a high level of interaction between the two directions. The single agent covers both the TX and RX directions. This single agent contains all of the monitors, BFMs, and sequence drivers required for both directions.

15.2.2 BFMs and Monitors

Monitors must be completely passive. The BFM does all of the activity of driving transactions. The BFM can make use of the monitor or duplicate some of the monitor's logic.

In general, most passive activity should be done by the monitor, while all active interactions with the DUT are done by the BFM. For example, the monitor might collect transactions and then emit an event for each transaction received. Upon this event, the BFM could be responsible for sending an acknowledgment back to the DUT.

In addition, the monitor rather than the BFM is used to collect transactions that come from the DUT. Some ways this can happen are as follows:

- In many protocols, each agent in the eVC has both a BFM and a monitor.

- In some protocols, the Receive side might not have any active components, so there is no BFM or sequence driver at all. In such cases, the Receive monitor can be placed inside a single agent that combines Receive and Transmit activities. This is especially convenient if the Receive and Transmit sides interact.

15.2.3 Clocks and Events

An eVC developer must decide whether to have the clock(s) propagated down to each agent or to use a central clock in the *env* to which all agents will refer.

The following guidelines are recommended:

- As far as possible, one centralized clock should be used, and it should be located in the *env*. (Similarly any other frequently occurring event should be centralized.) The main reason for this is to eliminate unnecessary performance overhead.

- On the other hand, if the protocol defines that agents can have different clocks (for example, different speeds), then each agent should have its own clock.

One should implement centralized clocks by maintaining a backpointer from the agents to the enclosing *env* **unit** and referring to events in the *env* as *env.clock*.

15.2.4 DUT Signals

It is important to differentiate DUT signals from other identifiers. References to DUT signals may be made by use of fields of type **string**. The names of these fields should have a **sig_** prefix. A separate prefix makes it simpler to distinguish these signals. The **sig_** fields should be placed in a natural location, typically either the *env* **unit** (if it is a signal used by all agents) or in an agent (for agent-specific signals).

15.2.5 Agent Details

Agents are the key to eVC architecture. In most eVCs, agents represent independent devices and have standard main elements. Some of the fields in the agents are also standard and should appear in all eVCs.

Agents are either *active* or *passive*. Active agents are agents that drive DUT signals. Passive agents never drive signals, either because they just monitor an interface within the DUT or because, according to the protocol, no signals need to be driven. Figure 15-4 shows the internals of a TX agent.

Figure 15-4 Agent Internals

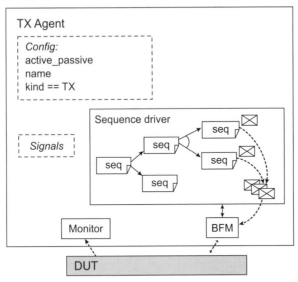

Passive agents represent DUT devices that are only monitored by the *e*VC. The passive agent therefore allows the projection of information inside the DUT onto the passive agent, which is a container of the information collected. Any of the agents in an *e*VC can be used as a passive agent. This allows for orderly monitoring of the DUT agent—collecting information and checking its behavior from a well-defined place in the *e*VC.

15.2.6 Combining *e*VCs

The earlier examples we looked at were standalone *e*VCs. We now look at a System-on-Chip (*XSoC*) example that combines several *e*VCs. Figure 15-5 shows such a verification environment.

Figure 15-5 XSoC eVC

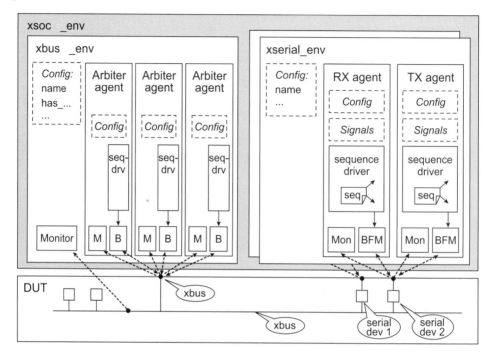

In this case, the *XSoC* verification environment could be the end user's full environment. Still, the *XSoC* verification environment can be represented as an *e*VC, because it highlights the fact that any current verification environment can turn into a verification component in a bigger verification environment in a future project. Designing a verification environment as an *e*VC helps create a consistent structure across different verification environments and projects.

In Figure 15-5, the *XSoC e*VC makes use of other *e*VCs. The *XSerial* and *XBus e*VCs are instantiated within the *XSoC e*VC. The *XSoC e*VC has no agents of its own. The DUT has two *XSerial* devices and one *XBus* device. Thus, the *XSoC e*VC has a list of *xserial_env* instances. The *XSoC e*VC integrates sub-*e*VC components to create a new and bigger verification component without adding much new code.

15.2.7 Typical Files in an *e*VC

Certain standard files are typically available with an *e*VC. Some files may be missing depending on the architecture of the *e*VC. The prefix *evc* for each file should be replaced by the name of the *e*VC.

Table 15-4 *e*VC Files

File Name	Description
*evc_*types.e	Global constants and types belonging to the *e*VC
evc_data_item.e	Definition of the data item struct *evc_data_item*
*evc_agent_*h.e	Header file: definition of an agent unit (optional)
evc_agent.e	Implementation of an agent unit
*evc_*env.e	Definition of the top-level unit evc_env
*evc_*monitor.e	Definition of central monitor unit, when relevant (for example, buses)
*evc_agent_*monitor.e	Definition of agent monitor unit, when relevant (for example, serial interface)
*evc_agent_*sequence.e	Predefined sequences / API methods
*evc_*checker.e	Protocol and data checks. Can be divided into agent/topic files
*evc_agent_*checker.e	Specific checks for each agent
*evc_*cover.e	Coverage definitions
*evc_agent_*cover.e	Coverage definitions
*evc_*wave.e	Definitions of wave commands for the *e*VC
*evc_*top.e	Imports all files Instantiates *e*VC entities Passed to **sn_compile.sh** (see **Chapter 16** for details)

15.3 *e*VC Example

This section examines the *e* code for the various components of the *XSerial e*VC. Note that we have skipped over the details of the *XSerial* Bus Protocol because this chapter focuses on the organization of the *XSerial e*VC. Note that *e* code for the high-level components is shown in this section. Some files are omitted in this book. Although all low-level details of the *XSerial e*VC may not be included in the *e* code, the purpose of this section is to give you an idea of structure of a typical *e*VC.

15.3.1 XSerial Bus Data Item

Example 15-1 describes the *e* code of data item for *XSerial e*VC.

Example 15-1 *e* Code for XSerial *e*VC Data Item

```
/*-------------------------------------------------------------------
File name    : vr_xserial_frame_payload_h.e
Title        : XSerial eVC frame payload structure
Project      : XSerial eVC
Developers   : Richard Vialls, Black Cat Electronics Ltd
Created      : 16-Apr-2002
Description  : his file declares the format of the generic XSerial frame
             : payload.
Notes        :
-------------------------------------------------------------------
Copyright 2002 (c) Verisity Design
-------------------------------------------------------------------*/
<'
package vr_xserial;

-- This type is used to enumerate the 2-bit frame_format field in the
-- frame. The various possible sub-types are extended in the files that
-- declare the sub-type formats. Note that the UNDEFINED value is used
-- in the case where the user wants to generate a payload with an
-- illegal frame_format value.
type vr_xserial_frame_format_t : [UNDEFINED = UNDEF];

-- This type is used to specify the status of a frame.
type vr_xserial_frame_status_t : [BAD_FORMAT]; -- illegal format field
                                               -- value
-- This struct contains the payload for a frame.
struct vr_xserial_frame_payload_s {

   -- This field indicates the status of the payload.
   status : list of vr_xserial_frame_status_t;
      keep soft status.size() == 0;

   -- This field holds the destination address for the frame.
   %destination : uint(bits:2);

   -- This field specifies the format of frame.
   frame_format : vr_xserial_frame_format_t;
      keep BAD_FORMAT in status => frame_format == UNDEFINED;  (continued...)
```

```
-- This field is the actual physical frame format field as encoded
-- in the frame.
%physical_frame_format : uint(bits:2);
    keep frame_format != UNDEFINED =>
        physical_frame_format == frame_format.as_a(uint);
    keep frame_format == UNDEFINED =>
        physical_frame_format not in
            all_values(vr_xserial_frame_format_t).all(it !=
                    UNDEFINED).apply(it.as_a(uint));
-- This method returns the payload as a list of bits. By using this
-- method rather than explicitly packing the struct, the user does
-- not need to be aware of the details of how a payload is packed.
-- Note that in the XSerial eVC example, this is trivial, but in
-- more complex eVCs, the process of packing a struct may be
-- complex.
pack_payload() : list of bit is {
    result = pack(packing.low, me);
}; -- pack_payload()

-- This method takes a bitstream and unpacks it into the payload
-- struct. As with pack_payload(), this method hides the
-- implementation details
-- of the struct packing/unpacking from the user.
unpack_payload(bitstream : list of bit, check_protocol : bool) is {

    -- Assume that this payload is going to be legal until we
    -- discover otherwise.
    status = {};

    -- The payload should always be exactly 12 bits long.
    if bitstream.size() != 12 {
        error("FATAL: Frame payload is not 12 bits long");
    };

    -- Unpack the destination and frame_format fields.
    unpack(packing.low, bitstream, destination,
            physical_frame_format);
    if physical_frame_format in
        all_values(vr_xserial_frame_format_t).apply(it.as_a(uint)) {
            frame_format =
                physical_frame_format.as_a(vr_xserial_frame_format_t);
    } else {
            frame_format = UNDEFINED;
    };

    -- Make the bitstream parameter point to the rest of the
    -- bitstream
    -- so that sub-typed extensions to this method can unpack the
    -- rest of the payload.
    bitstream = bitstream[4..];

    -- Make sure we've got a legal frame format.          (continued...)
```

```
        if check_protocol {
            check that frame_format != UNDEFINED
                else dut_error("Illegal frame format found: ",
                    physical_frame_format);
        };

        -- If the frame format is illegal, then set the status field
        -- accordingly.
        if frame_format == UNDEFINED {
            status.add(BAD_FORMAT);
        };

    }; -- unpack_payload()

    -- This method returns a convenient string representation of the
    -- contents of the payload. This is used for logging,
    -- waveform viewing, etc.
    nice_string(): string is {
        result = appendf("DEST:%01d", destination);
    }; -- nice_string()

    -- This method compares this payload with a payload supplied as a
    -- parameter. If the compare_dest field is false, then differences
    -- in the destination fields are ignored. It returns a list of
    -- strings that contains all detected differences.
    compare_payloads(exp_payload : vr_xserial_frame_payload_s,
                    compare_dest : bool) : list of string is {
        if compare_dest and exp_payload.destination != destination {
            result.add(append("Expected destination field: ",
                            bin(exp_payload.destination),
                            ", Actual destination field: ",
                            bin(destination)));
        };
        if exp_payload.frame_format != frame_format {
            result.add(append("Expected format field: ",
                            exp_payload.frame_format,
                            ", Actual format field: ",
                            frame_format));
            return result;
        };
    }; -- compare_payloads()

}; -- struct vr_xserial_frame_payload_s


-- If we have an illegal frame_format field, then none of the sub-typed
-- extensions will come into play so the remaining 8 bits of the
-- payload won't be there. We need the payload to always be 12 bits
-- long, so extend payloads with illegal frame formats to provide a
-- dummy 8-bit field.
extend UNDEFINED vr_xserial_frame_payload_s {
    -- This field pads out the remaining 8 bits if the frame format is
    -- illegal.                                              (continued...)
```

```
%dummy : byte;

-- Unpack the remaining bits into the dummy field
unpack_payload(bitstream : list of bit, check_protocol : bool) is also {
    unpack(packing.low, bitstream, dummy);
}; -- unpack_payload()

-- Make sure that if this payload gets printed, the bad frame format
-- is included in the string.
nice_string(): string is also {
    result = appendf("%s BAD_FORMAT:%02b ",
                     result,
                     physical_frame_format.as_a(uint(bits:2)));
}; -- nice_string()

}; -- extend UNDEFINED vr_xserial_frame_payload_s
'>
```

15.3.2 XSerial Bus Agent

Example 15-2 describes the *e* code of the agent for *XSerial e*VC.

Example 15-2 *e* Code for XSerial *e*VC Agent

```
/*----------------------------------------------------------------------------
File name    : vr_xserial_agent_h.e
Title        : Agent unit public interface
Project      : XSerial eVC
Developers   : Richard Vialls, Black Cat Electronics Ltd
Created      : 03-Jan-2002
Description  : This file declares the public interface of the agent unit.
Notes        : Because the agent handles both the TX and RX parts of a link,
             : the 'has_tx_path' and 'has_rx_path' fields are used to
             : sub-type the agent so that the TX and RX paths can be
             : separately disabled. These fields are controlled by the user
             : accesible field 'directions'
------------------------------------------------------------------------------
Copyright 2002 (c) Verisity Design
----------------------------------------------------------------------------*/

<'

package vr_xserial;

-- This type enumerates the possible modes of operation of an agent regarding
-- data directions.
type vr_xserial_directions_t : [TX_AND_RX, TX_ONLY, RX_ONLY];


-- This unit represents the eVCs main agent. An agent handles one
-- bidirectional XSerial link.
unit vr_xserial_agent_u {

    -- This field holds the logical name of the eVC instance this agent is
    -- contained in
    name : vr_xserial_env_name_t;

    -- This field holds the signal name of the TX_CLOCK signal on the XSERIAL
    -- interface.
    sig_tx_clock : string;

    -- This field holds the signal name of the TX_DATA signal on the XSERIAL
    -- interface.                                                   (continued...)
```

```
sig_tx_data  : string;

-- This field holds the signal name of the RX_CLOCK signal on the XSERIAL
-- interface.
sig_rx_clock : string;

-- This field holds the signal name of the RX_DATA signal on the XSERIAL
-- interface.
sig_rx_data  : string;

-- This field holds the signal name of the RESET signal on the XSERIAL
-- interface. If there is no reset signal in the design, then this
-- field should be left unconstrained.
sig_reset : string;
   keep soft sig_reset == "";

-- This field controls the active level of the RESET signal. By default,
-- the reset is active high, but by constraining this field to 0, it
-- can be set to active low.
reset_active_level : bit;
   keep soft reset_active_level == 1;

-- This field determines what reset state the eVC starts in at the
-- beginning of a test. By default, the eVC assumes that reset is
-- asserted at the start of a test. Where the sig_reset field is
-- constrained, the reset_asserted field will then go to FALSE at the
-- first de-assertion of the reset signal. Where the sig_reset field
-- is not constrained, the eVC can then be brought out of the reset state
-- by emitting the reset_end event.
reset_asserted : bool;
   keep soft reset_asserted == TRUE;

-- This field specifies whether this agent is ACTIVE or PASSIVE. In ACTIVE
-- mode, the agent drives signals and models one end of an XSerial link.
-- In PASSIVE mode the agent never drives signals and just monitors both
-- directions of an XSerial link.
active_passive : erm_active_passive_t;
   keep soft active_passive == ACTIVE;

-- This field allows the user to disable either the TX path or the RX
-- path. This can be used in cases where testing is only required of
-- one direction. When this field is TX_ONLY, flow control cannot work
-- and is disabled (i.e. the eVC will always consider the TX link as
-- ready).
directions : vr_xserial_directions_t;
   keep soft directions == TX_AND_RX;

-- If this field is TRUE then the agent does protocol checking on the
-- tx_data signal.
check_tx_protocol : bool;
   keep soft check_tx_protocol == TRUE;

-- If this field is TRUE then the agent does protocol checking on the
-- rx_data signal.
check_rx_protocol : bool;
   keep soft check_rx_protocol == TRUE;

-- This field controls the period of the TX clock in simulator time units.
-- If this field is 0, then the eVC does not generate the clock. Note that
-- this field is only used if the agent is ACTIVE and has a tx path. It
-- is recommended that this field be constrained using physical time units
-- e.g.: keep tx_clock_period == 20 ns; This ensures that there is no
-- dependency on the simulator time resolution.
tx_clock_period : time;
   keep soft tx_clock_period == 0;
```

(continued...)

```
    -- If this field is not "" then the agent writes a log file of that
    -- name with a .elog extension. This log file contains all TX
    -- transactions. If both this field and rx_log_filename are the
    -- same then both TX and RX log information will be written to a
    -- single file.
    tx_log_filename : string;
        keep soft tx_log_filename == "vr_xserial";

    -- If this field is not "" then the agent writes a log file of that
    -- name with a .elog extension. This log file contains all RX
    -- transactions. If both this field and tx_log_filename are the
    -- same then both TX and RX log information will be written to a
    -- single file.
    rx_log_filename : string;
        keep soft rx_log_filename == "vr_xserial";

    -- The user can extend this method to hook in a scoreboard. The user MUST NOT
    -- alter the frame in any way.
    tx_frame_completed(frame : vr_xserial_frame_s) is empty;

    -- The user can extend this method to hook in a scoreboard. The user MUST NOT
    -- alter the frame in any way.
    rx_frame_completed(frame : vr_xserial_frame_s) is empty;

    -- This field is set to the frame just transmitted just before the
    -- tx_frame_done event is emitted.
    !tx_frame : MONITOR vr_xserial_frame_s;

    -- This field is set to the frame just received just before the
    -- rx_frame_done event is emitted.
    !rx_frame : MONITOR vr_xserial_frame_s;

    -- This event gets emitted each time a frame is transmitted. In
    -- conjunction with the tx_frame field, it can be used as an
    -- alternative scoreboard hook to tx_frame_completed().
    event tx_frame_done;

    -- This event gets emitted each time a frame is transmitted. In
    -- conjunction with the rx_frame field, it can be used as an
    -- alternative scoreboard hook to rx_frame_completed().
    event rx_frame_done;

    -- This event gets emitted when reset is asserted. If there is no reset
    -- signal, then the user can emit this event directly to initiate reset
    -- for this instance of the eVC.
    event reset_start;

    -- This event gets emitted when reset is de-asserted. If there is no
    -- reset signal, then the user can emit this event directly to terminate
    -- reset for this instance of the eVC.
    event reset_end;

    -- This field is used to sub-type the agent for when the TX path is
    -- enabled.
    has_tx_path : bool;
        keep has_tx_path == value(directions in [TX_ONLY, TX_AND_RX]);

    -- This field is used to sub-type the agent for when the RX path is
    -- enabled.
    has_rx_path : bool;
        keep has_rx_path == value(directions in [RX_ONLY, TX_AND_RX]);

}; -- unit vr_xserial_agent_u


extend vr_xserial_monitor_u {                              (continued...)
```

```
    -- This field is a back-pointer to the agent that contains the monitor.
    agent : vr_xserial_agent_u;

}; -- extend vr_xserial_monitor_u


extend vr_xserial_tx_bfm_u {

    -- This field is a back-pointer to the agent that contains the BFM.
    agent : vr_xserial_agent_u;

}; -- extend vr_xserial_tx_bfm_u


extend has_tx_path vr_xserial_agent_u {

    -- This field is the instance of the monitor for the transmit direction.
    -- It only exists if the TX direction is enabled for this agent.
    tx_monitor : TX vr_xserial_monitor_u is instance;
        keep tx_monitor.file_logger.to_file == value(tx_log_filename);
        keep tx_monitor.sig_clock == value(sig_tx_clock);
        keep tx_monitor.sig_data == value(sig_tx_data);
        keep tx_monitor.agent == me;
        keep tx_monitor.has_protocol_checker == value(check_tx_protocol);

}; -- extend has_tx_path vr_xserial_agent_u


extend has_rx_path vr_xserial_agent_u {

    -- This field is the instance of the monitor for the receive direction.
    -- It only exists if the RX direction is enabled for this agent.
    rx_monitor : RX vr_xserial_monitor_u is instance;
        keep rx_monitor.file_logger.to_file == value(rx_log_filename);
        keep rx_monitor.sig_clock == value(sig_rx_clock);
        keep rx_monitor.sig_data == value(sig_rx_data);
        keep rx_monitor.agent == me;
        keep rx_monitor.has_protocol_checker == value(check_rx_protocol);

}; -- extend has_rx_path vr_xserial_agent_u


-- If in ACTIVE mode with TX path enabled, then add a TX BFM and a TX sequence
-- driver.
extend ACTIVE has_tx_path vr_xserial_agent_u {

    -- This field is the instance of the transmit BFM. It only exists if the
    -- TX direction is enabled for this agent and the agent is ACTIVE.
    tx_bfm : vr_xserial_tx_bfm_u is instance;
        keep tx_bfm.agent == me;
        keep tx_bfm.driver == value(tx_driver);
        keep tx_bfm.tx_monitor == value(tx_monitor);
        keep tx_bfm.sig_tx_clock == value(sig_tx_clock);
        keep tx_bfm.sig_tx_data == value(sig_tx_data);

    -- This field is the instance of the transmit sequence driver. It only
    -- exists if the TX direction is enabled for this agent and the agent is
    -- ACTIVE.
    tx_driver : vr_xserial_tx_driver_u is instance;
        keep tx_driver.name == value(name);

}; -- extend ACTIVE has_tx_path vr_xserial_agent_u

-- If both monitors are present, make them aware of each other.          (continued...)
```

```
extend has_rx_path has_tx_path vr_xserial_agent_u {

    -- Note that these pointers need to be set up in post_generate to avoid
    -- a generation order cycle.
    post_generate() is also {
        tx_monitor.reverse_monitor = rx_monitor;
        rx_monitor.reverse_monitor = tx_monitor;
    }; -- post_generate()

}; -- extend has_rx_path has_tx_path vr_xserial_agent_u


-- If agent is in ACTIVE mode and both paths are enabled, then the Tx BFM needs
-- to know where the Rx monitor is so it can check on the current state of
-- flow control
extend ACTIVE has_rx_path has_tx_path vr_xserial_agent_u {

    keep tx_bfm.rx_monitor == value(rx_monitor);

}; -- extend ACTIVE has_rx_path has_tx_path vr_xserial_agent_u

'>
```

15.3.3   XSerial Bus Tx BFM

Example 15-3 describes the *e* code for Tx BFM of *XSerial e*VC.

Example 15-3 *e* Code for XSerial *e*VC Tx BFM

```
/*-------------------------------------------------------------------------
File name    : vr_xserial_bfm_tx_h.e
Title        : TX BFM public interface
Project      : XSerial eVC
Developers   : Richard Vialls, Black Cat Electronics Ltd
Created       : 23-Jul-2002
Description : This file declares the public interface of the TX BFM unit.
Notes        : There is no RX BFM because the RX link direction contains no
             : signals that the eVC can drive. As such, the entire RX path
             : functionality is contained in the RX monitor and the eVC does
             : not need either an RX BFM or an RX sequence driver.
             :
             : The BFM depends on the functionality of the RX and TX
             : monitors.
-------------------------------------------------------------------------
Copyright 2002 (c) Verisity Design
-------------------------------------------------------------------------*/

<'

package vr_xserial;


-- This unit is the BFM used for the transmit direction (i.e. sending data
-- from the eVC to the DUT).
unit vr_xserial_tx_bfm_u {

    -- This field is a pointer to the TX monitor.
    tx_monitor : vr_xserial_monitor_u;

    -- This field is a pointer to the RX monitor (or NULL if the RX path is
    -- disabled).
    rx_monitor : vr_xserial_monitor_u;
        keep soft rx_monitor == NULL;

    -- This field is a pointer to the sequence driver.
driver : vr_xserial_tx_driver_u;                          (continued...)
```

```
    -- This field holds the signal name of the TX_CLOCK signal on the XSERIAL
    -- interface.
    sig_tx_clock : string;

    -- This field holds the signal name of the TX_DATA signal on the XSERIAL
    -- interface.
    sig_tx_data  : string;

}; -- unit vr_xserial_tx_bfm_u


'>
```

15.3.4 XSerial Bus Agent Tx Sequence

Example 15-4 describes the *e* code of the agent Tx sequence for *XSerial e*VC.

Example 15-4 *e* Code for XSerial *e*VC Agent Tx Sequence

```
/*-------------------------------------------------------------------------
File name   : vr_xserial_sequence_tx_h.e
Title       : Sequence stuff for TX side
Project     : XSerial eVC
Developers  : Richard Vialls, Black Cat Electronics Ltd
Created     : 07-Nov-2001
Description : This file implements TX sequences.
Notes       :
-------------------------------------------------------------------------
Copyright 2002 (c) Verisity Design
-------------------------------------------------------------------------*/

<'

package vr_xserial;

-- This type enumerates the logical names of each instance of the eVC in the
-- Verification Environment.
type vr_xserial_env_name_t : [];


-- This creates a sub-type of the frame struct that has additional
-- information relating to generation of frames by the sequence
-- interface.
extend TX vr_xserial_frame_s {

    -- This field can be used to sub-type the TX frame according to the
    -- eVC instance that is sending it.
    name : vr_xserial_env_name_t;
        keep name == value(driver.name);

    -- This field controls the delay before transmission of this frame in
    -- clock cycles - timed from the end of the previous frame.
    transmit_delay : uint;
        keep soft transmit_delay in [5..20];

}; -- extend TX vr_xserial_frame_s


-- This is the generic sequence struct for transmitting frames from the eVC.
sequence vr_xserial_tx_sequence using
    item=TX vr_xserial_frame_s,
    created_driver=vr_xserial_tx_driver_u;


-- Provide extensions to the sequence driver so that the driver, the sequence (continued...)
```

```
-- and the sequence items can all be sub-typed by the instance name of the
-- eVC.
extend vr_xserial_tx_driver_u {

    -- This field holds the name of the env this driver is contained in.
    name : vr_xserial_env_name_t;

    -- propagate master name to sequence tree
    keep sequence.name == value(name);

}; -- extend vr_xserial_tx_driver_u


extend vr_xserial_tx_sequence {

    -- This field allows sequences to be sub-typed on the name of the env.
    name : vr_xserial_env_name_t;
        keep name == value(driver.name);

    -- This is a utility field for basic sequences. This allows the user to
    -- do "do frame ...".
    !frame: TX vr_xserial_frame_s;

}; -- extend vr_xserial_tx_sequence


'>
```

15.3.5 XSerial Bus Monitor

Example 15-5 describes the *e* code of the monitor for *XSerial e*VC.

Example 15-5 *e* Code for XSerial *e*VC Monitor

```
/*------------------------------------------------------------------------
File name    : vr_xserial_monitor_h.e
Title        : XSerial monitor unit public interface
Project      : XSerial eVC
Developers   : Richard Vialls, Black Cat Electronics Ltd
Created      : 23-Jul-2002
Description  : The monitor unit implements a monitor for one direction of
             : traffic on an XSerial interface. This file defines the public
             : interface of the monitor unit. One instance of the monitor
             : unit is placed in the agent for each data direction. The
             : monitor is largely self-contained and standalone. It creates
             : instances of the MONITOR sub-type of the frame (which
             : contains additional information to the GENERIC sub-type) and
             : optionally writes information to a log file.
Notes        :
------------------------------------------------------------------------
Copyright 2002 (c) Verisity Design
------------------------------------------------------------------------*/

<'

package vr_xserial;


-- This type is used by the monitor to indicate which direction of data
-- transfer the monitor is monitoring.
type vr_xserial_direction_t : [TX, RX];


-- Provide a message tag that can be used to direct certain message
-- actions to a log file.
extend message_tag: [VR_XSERIAL_FILE];                          (continued...)
```

```
-- Provide a MONITOR frame sub-type with some extra fields and methods to
-- assist in monitoring.
extend MONITOR vr_xserial_frame_s{

    -- This field indicates which eVC instance this frame is associated with.
    name : vr_xserial_env_name_t;

    -- This field holds the time between the end of the previous frame and
    -- the beginning of this one.
    delay : uint;

}; -- extend MONITOR vr_xserial_frame_s


-- This unit contains a monitor that is capable of monitoring traffic in
-- one direction on an XSerial link. Two instances of this unit are required
-- to monitor a bidirectional link.
unit vr_xserial_monitor_u {

    -- This field indicates whether the monitor is monitoring the TX or RX
    -- direction.
    direction : vr_xserial_direction_t;

    -- This is the logger used for creating log files.
    file_logger: message_logger;
        keep file_logger.tags == {VR_XSERIAL_FILE};
        keep soft file_logger.to_screen == FALSE;
        keep soft file_logger.format == none;
        keep soft file_logger.verbosity == FULL;

    -- This is the clock signal of the link being monitored.
    sig_clock : string;

    -- This is the data signal of the link being monitored.
    sig_data : string;

    -- If this field is TRUE then the agent does protocol checking
    has_protocol_checker : bool;

    -- This event is the clock rise event for this direction prior to
    -- qualification with reset.
    event unqualified_clock_rise is rise('(sig_clock)') @sim;

    -- This event is the clock fall event for this direction prior to
    -- qualification with reset.
    event unqualified_clock_fall is rise('(sig_clock)') @sim;

    -- This is the main clock rise event for this direction. Note that this
    -- clock is qualified with reset and so is only emitted when reset is
    -- not asserted.
    event clock_rise;

    -- This is the main clock fall event for this direction. Note that this
    -- clock is qualified with reset and so is only emitted when reset is
    -- not asserted.
    event clock_fall;

    -- This field is a pointer to the monitor for the reverse link direction.
    -- If such a monitor does not exist (perhaps one link direction has been
    -- disabled), then this field will be NULL.
    reverse_monitor : vr_xserial_monitor_u;
        keep soft reverse_monitor == NULL;

    -- This field indicates whether the device sending the frames that this  (continued...)
```

```
-- monitor is monitoring is currently ready to receive frames or not.
!ready : bool;

-- The monitored frame is built up in this field.
!monitor_frame : MONITOR vr_xserial_frame_s;

-- This field counts the number of frames this monitor detects over the
-- duration of the test.
num_frames : uint;
    keep num_frames == 0;

-- This event is emitted at the start of a frame.
event frame_started;

-- This event is emitted at the end of a frame.
event frame_ended;

-- This TCM starts the monitor. It gets called by the agent to start the
-- monitor at the start of the test and each time reset is asserted. The
-- user can delay activation of the monitor by extending this method using
-- 'is first'.
start_tcms() @sys.any is undefined;

}; -- unit vr_xserial_monitor_u

'>
```

15.3.6 XSerial Bus Checker

Example 15-6 describes the *e* code of the checker for *XSerial e*VC.

Example 15-6 *e* Code for XSerial *e*VC Checker

```
/*----------------------------------------------------------------------
File name   : vr_xserial_protocol_checker.e
Title       : Protocol checker
Project     : XSerial eVC
Developers  : Richard Vialls, Black Cat Electronics Ltd
Created     : 03-Jan-2002
Description : This file contains the optional protocol checker
            : functionality within the monitor.
Notes       :
----------------------------------------------------------------------
Copyright 2002 (c) Verisity Design
----------------------------------------------------------------------*/

<'

package vr_xserial;

extend has_protocol_checker vr_xserial_monitor_u {

    -- This event is emitted in each clock cycle where the remote device
    -- has signalled that it is not ready to receive data frames. The falling
    -- edge of clock is used to ensure that there are no race conditions in
    -- the case where both monitors are clocked off the same clock signal. If
    -- the rising edge is used then the order in which the RX and TX monitors
    -- react to the rising edge of clock can affect the protocol checker.
    event halted is true((reverse_monitor != NULL) and
                         (not reverse_monitor.ready)) @clock_fall;

    -- This event is the same as halted but delayed to the next rising edge of
    -- clock.
    event halted_del is {@halted; [1]} @clock_rise;

    -- This event is emitted each time a frame starts while the remote device (continued...)
```

```
    -- is not ready. Note that halted_del is used to allow for the
    -- possibility that the frame started at the same clock edge as a HALT
    -- frame was received (which is allowed).
    event frame_start_halted is (@halted_del and @frame_started) @clock_rise;

    -- This event is emitted each time a message frame ends
    event message_frame_ended is
        true(monitor_frame.payload is a MESSAGE vr_xserial_frame_payload_s) @frame_ended;

    -- At the end of each frame, check that either it was a message frame or
    -- the remote device hadn't signalled a HALT before the start of the
    -- frame. Only message frames are allowed to start while the remote
    -- device is not ready.
    expect exp_send_while_not_ready is
        (not @frame_start_halted) or @message_frame_ended @frame_ended
        else dut_error("Non-message frame sent while remote device was not ready");

}; -- extend has_protocol_checker vr_xserial_monitor_u


-- Whenever a DUT error occurs in the monitor, print some useful information
-- about which monitor it occurred in.
extend dut_error_struct {

    write() is first {

        -- Ascertain which unit originated the error
        var caller_unit : any_unit = source_struct().get_unit();

        -- depending on what sort of unit originated the error, amend the error message
        if caller_unit is a vr_xserial_monitor_u (monitor) {
            message = append(monitor.agent.err_header(), ", ", monitor.err_header(),
                    ":\n", message);
        };

    }; -- write()

}; -- extend dut_error_struct

'>
```

15.3.7 XSerial Bus Coverage

Example 15-7 describes the *e* code for coverage of *XSerial e*VC.

Example 15-7 *e* Code for XSerial *e*VC Coverage

```
/*------------------------------------------------------------------------
File name   : vr_xserial_coverage_h.e
Title       : Coverage definitions
Project     : XSerial eVC
Developers  : Richard Vialls, Black Cat Electronics Ltd
Created     : 08-Jan-2002
Description : This file provides a basic functional coverage model for
            : frames. The user can extend this to create additional
            : coverage as required.
Notes       : Two different coverage events are used here because, at
            : present, Specman Elite only supports one-dimensional per-instance
            : coverage. In this case, we have two dimensions: the agent
            : name and the direction. By using two events, one for each
            : direction, separate coverage groups can be maintained for
            : each direction.
------------------------------------------------------------------------
Copyright 2002 (c) Verisity Design
------------------------------------------------------------------------*/
```

```
<'

package vr_xserial;

extend vr_xserial_frame_s {

    -- This event is emitted by the appropriate TX monitor whenever a frame
    -- transmission completes. It is used to collect coverage.
    event tx_frame_done;

    -- This event is emitted by the appropriate RX monitor whenever a frame
    -- reception completes. It is used to collect coverage.
    event rx_frame_done;

}; -- extend vr_xserial_frame_s


extend vr_xserial_monitor_u {

    -- Each time a frame ends, emit the event used to collect coverage for
    -- that frame.
    on frame_ended {
        case direction {
            TX : { emit monitor_frame.tx_frame_done; };
            RX : { emit monitor_frame.rx_frame_done; };
        };
    };

}; -- extend vr_xserial_monitor_u


extend MONITOR vr_xserial_frame_s {

    cover tx_frame_done is {
        item name using per_instance;
        item destination : uint(bits:2) = payload.destination;
        item frame_format : vr_xserial_frame_format_t = payload.frame_format;
        item data : byte = payload.as_a(DATA vr_xserial_frame_payload_s).data
            using when = (payload is a DATA vr_xserial_frame_payload_s);
        item message : vr_xserial_frame_message_t =
            payload.as_a(MESSAGE vr_xserial_frame_payload_s).message
            using when = (payload is a MESSAGE vr_xserial_frame_payload_s);
        item parity;
        item delay;
    }; -- cover tx_frame_done

    cover rx_frame_done is {
        item name using per_instance;
        item destination : uint(bits:2) = payload.destination;
        item frame_format : vr_xserial_frame_format_t = payload.frame_format;
        item data : byte = payload.as_a(DATA vr_xserial_frame_payload_s).data
            using when = (payload is a DATA vr_xserial_frame_payload_s);
        item message : vr_xserial_frame_message_t =
            payload.as_a(MESSAGE vr_xserial_frame_payload_s).message
            using when = (payload is a MESSAGE vr_xserial_frame_payload_s);
        item parity;
        item delay;
    }; -- cover rx_frame_done

}; -- extend vr_xserial_frame_s

'>
```

(continued...)

15.3.8 XSerial Bus Env

Example 15-8 describes the *e* code for the environment of *XSerial e*VC.

Example 15-8 *e* Code for XSerial *e*VC Env

```
/*-------------------------------------------------------------------------
File name    : vr_xserial_env_h.e
Title        : Env unit public interface
Project      : XSerial eVC
Developers   : Richard Vialls, Black Cat Electronics Ltd
Created      : 03-Jan-2002
Description  : This file contains the declaration of the env unit and all
             : user accessible fields, events and methods.
Notes        : In this eVC, there is only a single agent per env. An env
             : should contain all agents that logically belong to the same
             : "network". In the case of the XSerial protocol, a network
             : consists only of two devices linked bidirectionally. Because
             : all the behaviour of one device can be deduced by looking at
             : the signals at the other device, there is no need to have a
             : passive agent monitoring the other end of the link. As such,
             : only a single agent is ever required per env.
-------------------------------------------------------------------------
Copyright 2002 (c) Verisity Design
-------------------------------------------------------------------------*/

<'

package vr_xserial;

-- This unit is the env - the top level unit of the eVC.
unit vr_xserial_env_u like any_env {

    -- This field provides a screen logger for the env. Note that this eVC
    -- has only a single agent per env, so there is no point adding a separate
    -- screen logger for each agent.
    logger: message_logger;
        keep soft logger.verbosity == NONE;

    -- This field holds the logical name of this eVC instance.
    name : vr_xserial_env_name_t;

    -- The short_name() method should return the name of this eVC instance.
    short_name(): string is {
        result = append(name);
    };

    -- This controls what colour the short name is shown in.
    short_name_style(): vt_style is {
        result = DARK_CYAN;
    };

    -- This field is the instance of the agent. Note that each env (eVC
    -- instance) has just one agent that handles both Tx and Rx directions.
    agent : vr_xserial_agent_u is instance;
        keep agent.name == value(name);

}; -- unit vr_xserial_env_u

'>
```

15.3.9 XSerial Bus Top

Example 15-9 describes the *e* code for top file of *XSerial e*VC.

Example 15-9 *e* Code for XSerial *e*VC Top

```
/*------------------------------------------------------------------------
File name    : vr_xserial_top.e
Title        : Top level of XSerial eVC
Project      : XSerial eVC
Developers   : Richard Vialls, Black Cat Electronics Ltd
Created      : 07-Nov-2001
Description  : This file imports all files in the eVC
Notes        : Files ending in _h.e are public files for the user to read
             : all other files could potentially be encrypted.
------------------------------------------------------------------------
Copyright 2002 (c) Verisity Design
------------------------------------------------------------------------*/

<'

package vr_xserial;

define VR_XSERIAL_VERSION_1_0_OR_LATER;
define VR_XSERIAL_VERSION_1_1_OR_LATER;
define VR_XSERIAL_VERSION_1_2_OR_LATER;
define VR_XSERIAL_VERSION_1_3_OR_LATER;
define VR_XSERIAL_VERSION_1_4_OR_LATER;
define VR_XSERIAL_VERSION_1_4;

import evc_util/e/evc_util_top;

import vr_xserial/e/vr_xserial_frame_payload_h;
import vr_xserial/e/vr_xserial_frame_h;
import vr_xserial/e/vr_xserial_frame_payload_data_h;
import vr_xserial/e/vr_xserial_frame_payload_message_h;
import vr_xserial/e/vr_xserial_sequence_tx_h;
import vr_xserial/e/vr_xserial_monitor_h;
import vr_xserial/e/vr_xserial_bfm_tx_h;
import vr_xserial/e/vr_xserial_agent_h;
import vr_xserial/e/vr_xserial_env_h;
import vr_xserial/e/vr_xserial_coverage_h;
import vr_xserial/e/vr_xserial_bfm_tx;
import vr_xserial/e/vr_xserial_monitor;
import vr_xserial/e/vr_xserial_agent;
import vr_xserial/e/vr_xserial_protocol_checker;
import vr_xserial/e/vr_xserial_end_test_h;
import vr_xserial/e/vr_xserial_package;
import vr_xserial/e/vr_xserial_scoreboard_h;

'>
```

15.4 Summary

- An *e*VC is an *e* Verification Component. It is a ready-to-use, configurable verification environment, typically focusing on a specific protocol or architecture (such as Ethernet, AHB, PCI, or USB).

- Each *e*VC consists of a complete set of elements for stimulating, checking, and collecting coverage information for a specific protocol or architecture. You can apply the *e*VC to your device under test (DUT) to verify your implementation of the *e*VC protocol or architecture.

- You can use an *e*VC as a full verification environment or add it to a larger environment. The *e*VC interface is viewable and hence can be the basis for user extensions. It is recommended that such extensions be done in a separate file. Maintaining the *e*VC in its original form facilitates possible upgrades.

- An *e*VC implementation is often partially encrypted, especially in commercial *e*VCs where authors want to protect their intellectual property. Most commercial *e*VCs require a specific feature license to enable them.

- In this chapter, we take a sample *XSerial* *e*VC to show how a typical *e*VC should be constructed. The *XSerial* *e*VC is an example of how to code a general-purpose *e*VC for a point-to-point protocol.

- Certain standard files are typically available with an *e*VC. Some files may be missing depending on the architecture of the *e*VC.

CHAPTER 16

Interfacing with C

Often, verification components exist in C or C++. These components are usually reference models that contain the algorithms required to simulate the behavior of the DUT or other elements of the system. Sometimes, there are legacy C or C++ models that must be integrated with existing verification methodology. This chapter discusses the methodology to integrate C/C++ models in Specman Elite.

Chapter Objectives

- Describe C interface features.
- Define the guidelines for using the C Interface.
- Describe the implementation of the C Interface.
- Understand how to call C routines from *e*.
- Explain how to call *e* routines from C.
- Describe compilation and linking of C files.
- Understand how to link with C++.

16.1 C Interface Features

The C interface has the following features:

- You can pass data between *e* modules and C modules.
- You can call C routines from *e* code, even if that C routine has no knowledge of *e*.
- You can use *e* data types in C code.
- You can call *e* methods and manipulate *e* data from C code.
- You can access any global data under **sys** in C code.
- For best performance, you can compile and statically link C code with Specman Elite.

- For easier debugging of your *e* code, you can dynamically load C code as a shared library into Specman Elite.[1]

16.2 Integrating C Files

For C routines to interact with *e* code, the C files have to be integrated with Specman Elite. A special script **sn_compile.sh** is provided with the Specman Elite installation to compile and link C code and *e* code in Specman Elite.

16.2.1 Integration Flow

To integrate C code and *e* code in Specman Elite perform the following steps:

1. Declare the C routines in *e* code using the **is C routine** keywords. These keywords in the *e* code indicate to the **sn_compile.sh** script that the particular routine or *e* method is implemented as a C routine. Therefore the computation of the return value for that routine or method is done in C code.

2. Often the C code needs access to *e* types or methods. Export the *e* types or methods used by C routines by adding **C export** statements in the *e* code. The **C export** statements in *e* code allow C code to access *e* types or methods.

3. Generate the C header files (*.h* files) from the *e* files using the **sn_compile.sh** script with the **-h_only** option. These C header files contain information declared in *e* code with **is C routine** keywords and **C export** statements. These header files are compiled in the C domain.

4. Include the generated header files in your C files using the **#include** C syntax. These header files provide information about accessing *e* types or methods from C files.

5. Compile all your C files using the **gcc** compiler (**gcc** with **-c** and **-o** options) to create object (*.o* files).

6. Compile your *e* files that contain C directives and link them with your simulator and C object files using the **sn_compile.sh** script with **-l** option and the **-o** option to create an executable binary file.

7. Execute this binary file at the operating system command prompt to run Specman Elite along with the compiled C code instead of running the standard Specman Elite binary.

Figure 16-1 below shows the integration mechanism. The exact commands to execute are shown later in Example 16-4.

---

1.   In this chapter, we do not discuss dynamic loading of C code. The dynamic linking mechanism is very similar to the static linking mechanism with minor differences. Therefore, the discussion of dynamic linking is left for Specman Elite reference manuals.

Figure 16-1 Integrating *e* Files and C Files

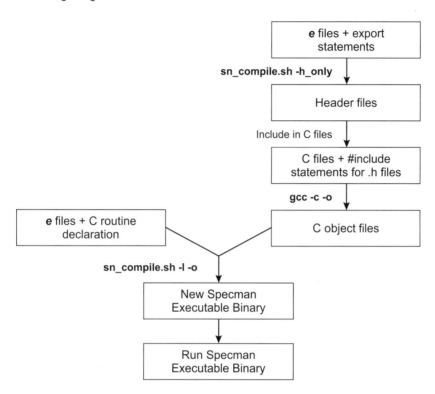

16.2.2 A Simple Example

A simple example described in this section shows how the integration of *e* files and C files takes place. This example uses a C routine, *wrap_mean_sqrt*, declared as a global *e* routine, *mean_sqrt()*, to calculate the mean square root of a list of packets.

16.2.2.1 Declaring a Global *e* Routine as a C Routine

First, a global *e* routine, *mean_sqrt()* must be declared. This routine *mean_sqrt()* is implemented using the C routine *wrap_mean_sqrt*. Example 16-1 shows the declaration of *mean_sqrt()* routine.

Example 16-1 Declaring a Global *e* Routine as a C Routine

```
File name: wrap_mean_sqrt.e
<'
-- Declare a global e routine named mean_sqrt which takes a list
-- of integers as an argument. The return type is also int.
-- The computation of this routine is performed in C code using
-- the C routine wrap_mean_sqrt

routine mean_sqrt(l:list of int):int is C routine wrap_mean_sqrt;

'>
```

16.2.2.2 Calling a Global *e* Routine Declared as a C Routine

The *e* routine *mean_sqrt()* declared earlier is invoked in *e* code simply by calling it like an *e* method. Example 16-2 calls *mean_sqrt()* and passes it a list of packet lengths. The *mean_sqrt()* calls the *wrap_mean_sqrt* C routine and passes it the list of packet lengths.

Example 16-2 Calling a Global *e* Routine Declared as a C Routine

```
File: packets_statistics.e
File calls the mean_sqrt e routine
to compute the mean square root of packet
lengths in a list. The mean_sqrt e routine
calls the C routine wrap_mean_sqrt.
<'
-- Definition of the packet struct
struct packet {
    kind       : [good,bad] ;
    addr       : uint (bits : 2) ;
    len        : uint (bits : 6) ;
    data [len] : list of byte ;
    !parity    : byte ;
};
```

Example 16-2 Calling a Global *e* Routine Declared as a C Routine (Continued)

```
-- Definition of port struct
struct port {
    packets         : list of packet; -- List of packets
    !len_mean_sqrt : int; -- Mean square root value
    !packets_lens   : list of uint (bits:6); -- List of lengths

    check() is also {
        -- Create a list of packet lengths
        packets_lens = packets.apply( it.len );
        -- Print the list of packet lengths
        print packets_lens;
        -- Call the mean_sqrt e routine with a list of packet
        -- lengths. This routine calls the C routine
        -- wrap_mean_sqrt to perform the computation.
        len_mean_sqrt = mean_sqrt( packets_lens );
        -- Print the computed value.
        print len_mean_sqrt;
    };
};

-- Instantiate the port struct
extend sys {
    port : port;
};
'>
```

16.2.2.3 Implementing the C Routine

The *wrap_mean_sqrt* C routine is implemented by means of another C routine *compute_mean_sqrt*. Therefore, there are three C files, *wrap_mean_sqrt.c*, and *compute_mean_sqrt.h*, *compute_mean_sqrt.c*. Example 16-3 shows the C code for all three files.

Example 16-3 Implement the C Routine

```
File name: wrap_mean_sqrt.c
#include "compute_mean_sqrt.h"
#include "wrap_mean_sqrt_.h"

int wrap_mean_sqrt( SN_LIST( int ) e_list_of_int ) {

    int i;
    int result;
    int list_size;
    double * c_list_of_double;
```

Example 16-3 Implement the C Routine (Continued)

```
    list_size = SN_LIST_SIZE( e_list_of_int );
    c_list_of_double =
        (double *)calloc( list_size, sizeof(double) );

    for (i=0; i < list_size; i++ ) {
        c_list_of_double[i] =
            (double)SN_LIST_GET( e_list_of_int, i, int);
    };

    result = (int)compute_mean_sqrt( c_list_of_double, list_size );
    free( c_list_of_double );
    return( result );

} // wrap_mean_sqrt
//End of file wrap_mean_sqrt.c
--------------------------------------------------------------------


--------------------------------------------------------------------
File name: compute_mean_sqrt.h
double compute_mean_sqrt( double *nums, int size );
//End of file compute_mean_sqrt.h
--------------------------------------------------------------------


--------------------------------------------------------------------
File name: compute_mean_sqrt.c
#include <math.h>
#include "compute_mean_sqrt.h"

double compute_mean_sqrt( double *nums, int size ) {
    int i;
    double sum = 0;
    for ( i=0; i<size; i++ ) {
        sum += sqrt( nums[ i ] );
    };
    return( sum/size );
};
//End of file compute_mean_sqrt.c
--------------------------------------------------------------------
```

16.2.2.4 Commands for Compiling and Linking *e* Code and C Code

Example 16-4 shows the commands required to compile and link *e* code and C code defined to implement the *mean_sqrt() e* routine. These commands are executed at the operating system prompt (for example the UNIX command prompt).

Example 16-4 Commands for Compiling and Linking *e* Code and C Code

```
This example does not contain e code.
It contains commands executed at the operating system prompt(%)

The file wrap_mean_sqrt.e contains the e routine declaration
for mean_sqrt that points to C routine wrap_mean_sqrt. The .h
file is generated using the sn_compile.sh -h_only option. The
following command generates the wrap_mean_sqrt_.h file.
    % sn_compile.sh -h_only wrap_mean_sqrt.e -o wrap_mean_sqrt_.h

The file compute_mean_sqrt.c performs the mean square root
computation. The file wrap_mean_sqrt.c contains an implementation of
the wrap_mean_sqrt C routine. This routine calls the compute_mean_sqrt
routine. The file wrap_mean_sqrt.c also includes the .h file
wrap_mean_sqrt_.h that was generated in the previous step. Compile the
files using the gcc -c -o option. These commands create the object
files.
    % gcc -c compute_mean_sqrt.c -o mean_sqrt.o
    % gcc -c wrap_mean_sqrt.c -o wrap_mean_sqrt.o

Compile the e files and C object files to create a new Specman Elite
binary file. Use the command sn_compile.sh -l. The -o option is used
to specify the name of the new Specman Elite binary called
float_mean_sqrt.
    % sn_compile.sh -l "mean_sqrt.o wrap_mean_sqrt.o" \
      wrap_mean_sqrt.e -o float_mean_sqrt

Run the new Specman Elite binary instead of running the standard
Specman Elite binary(specman).
    % ./float_mean_sqrt

This brings up the Specman prompt. The new binary that is executing
contains both standard Specman Elite behavior plus the C code.
Load the packet_statistics.e file and run the test using the test
command. In this environment, calls can be made to the mean_sqrt()
global e routine.
    Specman wrap_mean_sqrt> load packets_statistics.e
    Specman packets_statistics> test
```

16.3 Accessing the *e* Environment from C

This section discusses how to access the *e* environment, its objects, fields, variables, types, enumerated types, and other environment objects from C code.

16.3.1 SN_TYPE Macro

The **SN_TYPE** macro declares an *e* type in C. Use this macro whenever an *e* enumerated type or struct must be declared in C. The syntax for the **SN_TYPE** C construct is shown below. The *type_name* can be any *e* type.

```
SN_TYPE(type-name)
```

Example 16-5 shows usage of the SN_TYPE macro. This example shows the declaration of a C routine that receives a string argument and returns a boolean value.

Example 16-5 SN_TYPE Macro

```
SN_TYPE(bool) check_name (SN_TYPE(string) name) {
....
}
```

16.3.2 SN_LIST Macro

The **SN_LIST** macro declares an *e* list type in C. Use this macro whenever an *e* list must be declared in C. The syntax for the **SN_LIST** C construct is shown below. The *type_name* can be any *e* type.

```
SN_LIST(type-name)
```

Example 16-6 shows usage of the **SN_LIST** macro. This example shows the declaration of a C routine that receives a list of names and a string argument and returns a boolean value.

Example 16-6 SN_LIST Macro

```
SN_TYPE(bool) item_is_in_list(
    SN_LIST(string) list_of_names, /*List of names*/
    SN_TYPE(string) name
)
{
....
}
```

16.3.3 SN_ENUM Macro

Any enum defined in *e* can be used in C like any other *e* type using the **SN_TYPE** macro. An enumerated value name can be specified with the **SN_ENUM** macro. The syntax for the **SN_ENUM** macro is shown below. The *enum_type_name* can be any *e* type.

```
SN_ENUM(enum-type-name,value-name)
```

Example 16-7 shows usage of the **SN_ENUM** macro.

Example 16-7 SN_ENUM Macro

```
In e code:
<'
type color: [red, green, blue]; -- Define enumerated type in e
C export color; -- Export the enumerated type to C
'>
--------------------------------------------------------------------
In C code:

SN_TYPE(color) last_color; /*Define a variable of enum color*/
last_color = SN_ENUM(color,red); /*Call to SN_ENUM*/
```

16.3.4 SN_DEFINE Macro

The **SN_DEFINE** macro specifies an *e* defined name in C. Use this macro whenever an *e* defined name is to be used in C. The syntax for the **SN_DEFINE** C construct is shown below.

```
SN_DEFINE(defined-name)
```

Example 16-8 shows usage of the **SN_DEFINE** macro.

Example 16-8 SN_DEFINE Macro

```
In e:
<'
define NUM_OF_PORTS 5;
'>
-------------------------------------------------------------------
In C:
unsigned int num_of_ports = SN_DEFINE(NUM_OF_PORTS);
```

16.3.5 Passing Objects between *e* and C

- Data can be passed as parameters from *e* to C and vice versa.
- Data types **int**, **uint**, **enum**, and **bool** are passed by value (a copy is passed).
- Data types **struct**, **list**, and **string** are passed by pointer (by reference).

A struct in *e* is represented in C as a pointer to a C struct. You access its fields using the **->** operator. The name of the C struct field is the same as defined in *e*. A struct field whose name is a keyword in C (like **int** or **for**) is not accessible in C. Example 16-9 shows how to access *e* struct fields in C code.

Example 16-9 Accessing *e* Struct Fields in C

```
File Name: example.e
<'
struct cell { -- Define new struct
    data: int;
};

struct packet { -- Define new struct
    len: int; -- Length field
        keep len == 50; -- Constraint on len field
    cell: cell; -- Instantiate another struct
};
C export packet;

extend sys {
    packet :packet; -- Instantiate the packet
    run() is also {
        pkt(); -- Call an e routine that is implemented as a C routine.
    };
};
'>


<'
-- Define an e routine pkt() that points to C routine
-- my_c_func
routine pkt() is C routine my_c_func;
'>
```

Example 16-9 Accessing *e* Struct Fields in C (Continued)

```
File name: example_me.c
#include "e_ifc_file_.h"

/*Define the routine to implement the pkt() e routine*/
void my_c_func()
{
    SN_TYPE(packet) p_packet; /*Define an e struct type in C*/
    SN_TYPE(cell) p_cell; /*Define an e struct type in C*/
    p_packet = SN_SYS->packet; /*Access field packet of sys struct,
                              SN_SYS corresponds to the sys struct
                              in e*/
    p_cell = p_packet->cell; /*Access field cell of packet struct*/
    printf("\n PACKET_LEN = %d \n",p_packet->len); /*print field*/
}
```

16.4 Calling C Routines from *e*

There are three ways to call C routines from *e*:

- You can declare a global *e* routine to be implemented as a C routine. In this case, you can call the C routine directly from anywhere in your *e* code.

- You can declare a local *e* method to be implemented as a C routine. In this case, you define an *e* method within a struct and indicate that the method's body is implemented as a C routine. When you call the *e* method, the name of the enclosing struct instance is passed to the C routine and the C routine is executed. This way of calling C routines is useful when the C routine manipulates the fields of a particular struct instance.

- You can declare a local *e* method to be implemented as a dynamic C routine or foreign dynamic C routine. Dynamic C routines are not discussed in this chapter.

16.4.1 Global *e* Routine

A global *e* routine statement is declared with a specified name, parameters, and result type. The body of the global *e* routine is implemented by the specified C routine. The syntax for a global *e* routine is as follows:

```
routine e-routine-name(param, ...) [:result-type] [is C routine c-
routine-name];
```

If the optional "**is C routine** *c-routine-name*" is omitted, the assumed name of the C routine is the *e-routine-name*. A global *e* routine is a statement, not a struct member declaration. An example of a global *e* routine was discussed in "Integrating C Files" on page 318.

16.4.2 *e* Method using C Routine

A *e* method can be declared in such a way that when the *e* method is called, the C routine is executed. The syntax for defining an *e* method using a C routine is as follows:

```
e-method-name(param,...)[:result-type] is C routine c-routine-name ;
```

The construct **method ... is C routine** is a struct member, not a statement declaration unlike a global *e* routine. Example 16-10 shows the usage of an *e* method implemented with use of a C routine. This example shows how to call a C routine, *proprietary_encrypt*, to encrypt the *data* field of a struct *packet*.

Example 16-10 *e* Method using C Routine

```
File name: encrypt.c
#include <stdio.h>
#include "encrypt_.h"

SN_TYPE(byte) proprietary_encrypt( /*Define C routine*/
    SN_TYPE(packet) packet1,
    SN_TYPE(byte) data
)
    {
        return( data + 8);
    };
-----------------------------------------------------------------

                                              (continued...)
```

Example 16-10 *e* Method using C Routine (Continued)

```
File name: encrypt.e
<'
import packet;

struct packet {
    kind       : [good,bad];
    addr       : uint (bits : 2) ;
    len        : uint (bits : 6) ;
    data [len] : list of byte ;
    !parity    : byte ;

    -- Declare method encrypt of struct packet that is implemented
    -- using the C routine proprietary encrypt.
    encrypt( byte ):byte is C routine proprietary_encrypt;

    post_generate() is {
        out("Encrypting data: before -\n",data); -- unencrypted data
        for each (d) in data {
            data[index] = encrypt( d ); -- Call encrypt method
                                        -- like a regular e method
        };
        out("after -\n",data); -- Print the encrypted data
    };// post_generate
}; // packet

extend sys {
    packets: list of packet; -- Instantiate a list of packets
        keep packets.size() == 5; -- Constrain the size of list
};
'>
```

16.5 Calling *e* Methods from C

Calling an *e* method from C is done by means of the **SN_DISPATCH** macro. You can use **SN_DISPATCH** in C routines that have been called from *e* or in C routines called from other C code. The syntax for the **SN_DISPATCH** macro is as follows:

```
SN_DISPATCH(method-name, enclosing-struct, type-name, (enclosing-
struct, params,...))
```

The arguments for **SN_DISPATCH** macro are shown in Table 16-1.

Table 16-1 SN_DISPATCH Macro Arguments

method-name	The name of the method to be called, exactly as declared in *e*.
enclosing-struct	The instance of the struct that contains the method. If SN_DISPATCH is used in a C routine that is called from *e*, then the method's enclosing struct instance was passed implicitly to the C routine when the C routine was called.
type-name	The *e* type of the struct.
params, ...	Parameters passed to the called method (if any parameters are required). The parentheses around the parameters are required.

The *e* method must be exported from *e* to C with **C export** as shown below:

```
C export your_struct_type.your_method();
```

Example 16-11 shows use of an *e* method call from a C routine by means of the **SN_DISPATCH** macro.

Example 16-11 Calling *e* Methods from C using SN_DISPATCH Macro

```
e File name: print_method.e
<'
struct event_methods {
   print_message(i : int) is { -- e method definition
      outf("Message to print this number: %d\n", i);
   };
};

C export event_methods.print_reply(); -- Export the e method
'>
-----------------------------------------------------------------
```

Example 16-11 Calling *e* Methods from C using SN_DISPATCH Macro (Continued)

```
C File name: print_method.c
#include <stdio.h>
#include "print_method.h"

/*Define C routine c_message. Whenever this C routine is
called, it will in turn call the e routine print_message*/
void c_message(int i) {
    /*Declare a variable in C of struct type event_methods*/
    SN_TYPE(event_methods) c_em = SN_STRUCT_NEW(event_methods);
    /*Call the print_message e method with argument i from C*/
    SN_DISPATCH(print_message, c_em, event_methods, (c_em, i));
}
```

16.6 C Export

The **C export** statement exports the *e* declared type or method to the generated C header file when compiling the *e* modules using the **sn_compile.sh** script. When a struct is exported, it in turn exports all the types of its fields, recursively. Methods of a struct must be exported explicitly. No parameters should appear in the method export statement, even if the method has parameters associated with it. The export statement can appear in any *e* module (where the types or methods are known). The syntax for the **C export** statement is as follows:

```
C export type-name ;
C export type-name.method-name();
```

Example 16-12 shows a **C export** statement usage.

Example 16-12 C Export Statement Usage

```
This example shows the usage of C export statements.
When a struct is exported, all fields of the struct are
recursively exported for use in the C code. However,
methods of the struct must be explicitly exported.
<'
type color: [red, green, blue]; -- Enumerated type
struct packet { -- Struct packet
    add(num: int) is { -- Method of struct packet.
    ...
    };
    ...
};
```

Example 16-12 C Export Statement Usage (Continued)

```
C export packet; -- Export struct packet for use in C code.
C export packet.add(); -- Export method packet.add() for use in C
                        -- code.
C export color; -- Export enumerated type.
'>
```

16.7 Guidelines for Using the C Interface

- Define types and variables in *e*, then call C routines, passing all needed information through parameters.

- Do not use fields where the enum type is declared implicitly in the field declaration. The enum value names will not be accessible on the C side.

- Do not save an *e* instance pointer in C. The garbage collection mechanism will destroy the pointed instance. Use a field under **sys** instead.

- Struct instances created by the appropriate macros should NOT be freed (de-allocated) in C. Such struct instances will be de-allocated automatically by Specman Elite when they are no longer needed. You should not attempt to de-allocate them manually.

- Complex structs can be packed into a list of bytes or bits before they are passed to C. This can eliminate the need in C to access multiple fields. For example, this can be useful when oneis computing a Cyclic Redundancy Code (CRC) in C code on blocks of data.

The following features are not supported in Specman Elite.

- Long integers are not supported by the C interface.
- Lists with keys are not supported by the C interface.
- TCMs cannot be implemented by C routines.
- Subtypes (when inheritance, for example, small packet) cannot be used as type specifiers. Struct instances of subtypes can be passed and used as regular types' instances.

16.8 Linking with C++ Code

You can integrate C++ code with Specman Elite using the Specman Elite C interface. C++ modifies all the function names by addtion of the argument type and possibly the return type information to the function names. Thus you cannot directly link with C++ code.

There are two ways to overcome this:

- You can add an **extern "C"** declaration before the function prototype in the corresponding C++ source file (or, in the .h file, if it is included by the .C file). For example, replace the following function prototype

```
int   loadmem( int type, char *cmd );
```

with

```
extern "C" int   loadmem( int type, char *cmd );
```

- If you do not have the C++ source file, you must write a small C++ wrapper that has the **extern "C"** declaration. You must compile the C++ wrapper with the same C++ compiler you used to compile the original C++ code, since different compilers modify the function name differently.

After compiling the C++ code, you should be able to integrate it as if it were C code. When integrating C++ code with standalone Specman Elite, you may need to compile the Specman Elite main() function using the same C++ compiler that compiled your C++ code. You can choose your own C++ compiler.

16.9 Summary

- Specman Elite provides a lot of flexibility in the interaction between e code and C code.

- A special script **sn_compile.sh** is provided with the Specman Elite installation to compile and link C code and e code in Specman Elite.

- The **SN_TYPE** macro declares an e type in C. Use this macro whenever an e enumerated type or struct must be declared in C.

- The **SN_LIST** macro declares an e list type in C. Use this macro whenever an e list must be declared in C.

- An enumerated value name can be specified with the **SN_ENUM** macro.

- The **SN_DEFINE** macro specifies an e defined name in C. Use this macro whenever an e defined name is to be used in C.

- Data can be passed as parameters between e and C. Data of type **int**, **uint**, **enum**, and **bool** are passed by value (a copy is passed). Data of type **struct**, **list**, and **string** are passed by pointer (by reference).

- A struct in e is represented in C as a pointer to a C struct. You access its fields using the **->** operator. The name of the C struct field is the same as defined in e. A struct field whose name is a keyword in C (like **int** or **for**) is not accessible in C.

- There are three ways to call a C routine from e code—a global e routine to be implemented as a C routine, a local e method to be implemented as a C routine or a local e method to be implemented as a dynamic C routine or foreign dynamic C routine. The **is C routine** keywords are used to implement these *e* routines. Dynamic C routines are not discussed in this chapter.

• Calling an *e* method from C is done using the **SN_DISPATCH** macro. You can use **SN_DISPATCH** in C routines that have been called from *e* or in C routines called from other C code.

• The **C export** statement exports the *e* declared type or method to the generated C header file when the *e* modules are compiled with the **sn_compile.sh** script. When a struct is exported, it in turn exports all the types of its fields, recursively. Methods of a struct must be exported explicitly.

• You can integrate C++ code with Specman Elite using Specman Elite C interface by means of some special workaround techniques.

PART 5

Appendices

A   **Quick Reference Guide**
Quick reference for important *e* syntax

B   *e* **Tidbits**
History of *e*, *e* Resources, Verification Resources

Quick Reference Guide

This appendix contains the syntax and usage for selected *e* constructs.

Abbreviations:

arg - argument	inst - instance
bool - boolean	num - number
enum - enumerated	TCM - time-consuming method
expr - expression	TE - temporal expression

A.1 Predefined Types

bit   // unsigned integer with value 0 or 1 (default: 0)

byte   // unsigned integer in the range 0-255 (default: 0)

int   // 32-bit signed integer (default: 0)

uint   // 32-bit unsigned integer (default: 0)

int | uint (bits: n | bytes: n)   // n-bit or n-byte signed int or uint

bool   // one-bit boolean (0 = FALSE, 1 = TRUE) (default: FALSE)

list [(key: *field-name*)] of *type*
              // a list of elements of the specified type (default: empty)

string   // strings are enclosed in quotes: "my string" (default: NULL)

Type Conversion

expr = *expr*.as_a(*type*)

A.2 Statements

A.2.1   User-Defined Types

struct *struct-type* [like *base-struct-type*] { struct members };

unit *unit-type* [like *base-unit-type*] { unit members };

type *type-name* : [u]int (bits: n | bytes: n);   // defines a scalar type

type *type-name* : [*name* [=n], ...];   // defines an enumerated type

extend *type-name* : [*name* [=n], ...]; // extends an enumerated type

extend *struct-type|unit-type* { additional struct or unit members };
              // extends a struct or unit

A.3 Structs and Unit Members

fields	constraints	when conditions
methods and TCMs	cover groups	events
temporal struct|unit members		
preprocessor directives		

A.3.1 Fields

[!][%]*field-name* : *type*;   // ! = do not generate, % = physical field

field-name[*n*] : list of *type*;   // creates a list with n elements

field-name : *unit-type* is instance;   // for units only, not structs

A.3.2 Conditional Extensions using When

```
type enum-type: [name1, name2, ...];
struct|unit struct-type|unit-type {
    field-name : enum-type;
    when name1 struct-type|unit-type { additional members };
};
extend name1 struct-type|unit-type { ... };
```

A.3.3 Constraints

keep [soft] *bool-expr*;   // for example, keep field1 <= MY_MAX

keep [soft] *field-name* in [*range*];   // example: keep field1 in [0..256]

keep *bool-expr1* => *bool-expr2*;   // bool-expr1 implies bool-expr2

keep [soft] *field-name* in *list*;

keep *list*.is_all_iterations(*field-name*);

keep *list1*.is_a_permutation(*list2*);

keep for each (*item*) in *list* { [soft] *bool-expr*, ... };

keep soft *bool-expr* == select { *weight* : *value*; ... };

keep [soft] gen (*item-a*) before (*item-b*);

keep *gen-item*.reset_soft();   // ignore soft constraints on gen-item

keep *field-name*.hdl_path() == *"string"* ; //field-name is unit instance

A.3.4   Methods and TCMs

regular-method([*arg* : *type*, ...]) [: *return-type*] is { *action*; ... };

TCM([*arg* : *type*, ...]) [: *return-type*] @*event-name* is { *action*; ... };

A.3.5   Extending or Changing Methods and TCMs

method(*arg* : *type*, ...) [: *return-type*] is also|first|only { *action*; ... };

TCM(*arg* : *type*, ...) [: *return-type*] @*event-name* is also|first|only
     { *action*; ... };

A.3.6   Predefined Methods of All Structs and Units

run()	extract()	check()	finalize()
init()	pre_generate()	post_generate()	
copy()	do_print()	print_line()	quit()

A.4 Actions

A.4.1   Generation On the Fly

gen *gen-item* ;

gen *gen-item* keeping { [soft] *constraint-bool-expr* ; ... };

A.4.2   Conditional Procedures

if *bool-expr* [then] { *action*; ... }
[else if *bool-expr* [then] { *action*; ... }]
[else { *action*; ... }] ;

case { *bool-expr*[:] { *action*; ... } ; [default[:] { *action*; ... } ;] };

case *expr* { *value*[:] { *action*; ... } ; [default[:] { *action*; ... } ;] };

A.4.3   Loops

for *i* from *expr* [down] to *expr* [step *expr*] [do] { *action*; ... };

for each [*struct-type*] (*list-item*) [using index (*index-name*)]
     in [reverse] *list* [do] { *action*; ... };

for each [line] [(*line-name*)] in file *file-name* [do] {*action*; ... };

while *bool-expr* [do] { *action*; ... };

break;   // break the current loop

continue;   // go to the next iteration of the loop

A.4.4   Invoking Methods and TCMs

TCM2()@event-name is { *TCM1()*; *method()*;};   // calling methods

method1() is { *method2()*; *method3()*; };   // calling methods

method() is { start *TCM()*;};   // starting a TCM on a separate thread

Note: A TCM can only be *called* from another TCM. However, a TCM can be *started* from a regular method or from another TCM.

A.4.5   Checks

check that *bool-expr* [else dut_error(...)];

A.4.6   Variable Declarations and Assignments

var *var-name* : *type*;   // declare a variable

var-name = *expr* ;   // e.g. field-name=expr, var-name=method()

var *var-name* : = *value*;   // declare and assign a variable

A.4.7   Printing

print *expr*[,...] [using *print-options*] ;

print *struct-inst* ;

A.4.8   Predefined Routines

A.4.8.1   Deep Copy and Compare Routines

deep_copy(*expr* : struct-type) : struct-type

deep_compare[_physical](*inst1*: struct-type, *inst2*: struct-type, *max-diffs*: int): list of string

A.4.8.2   Output Routines

out ("*string*", *expr*, ...);   out (*struct-inst*);

outf ("*string %c ...*", *expr*); // c is a conversion code: s, d, x, b, o, u

A.4.8.3 Selected Configuration Routines

set_config(*category*, *option*, *option-value*)

get_config(*category*, *option*);

A.4.8.4 Selected Arithmetic Routines

min|max (x: int, y: int): int   abs(x: int): int

ipow(x: int, y: int): int   isqrt(x: int): int

odd|even (x: int): bool   div_round_up(x: int, y: int): int

A.4.8.5 Bitwise Routines

expr.bitwise_and|or|xor|nand|nor|xnor(*expr*: int|uint): bit

A.4.8.6 Selected String Routines

appendf(*format*, *expr*, ...): string   append(*expr*, ...): string

expr. to_string(): string   bin|dec|hex(*expr*, ...): string

str_join(*list*: list of string, *separator*: string): string

str_match(*str*: string, *regular-expr*: string): bool

str_replace(*str*:string, *regular-expr*:string, *replacement*:string):string

str_split(*str*: string, *regular-expr*: string): list of string

A.4.8.7 Selected Operating System Interface Routines

system("*command*"): int   date_time(): string

output_from("*command*"): list of string

output_from_check("*command*"): list of string

get_symbol(*UNIX-environment-variable*: string) : string

files.write_string_list(*file-name*: string, *list*: list of string)

A.4.8.8 Stopping a Test

stop_run();   // stops the simulator and invokes test finalization

A.5 Operators

Operator precedence is left to right, top to bottom in the list

[ ] list indexing	[..] list slicing
[:] bit slicing	$f()$ method or routine call
. field selection	in range list
{... ; ...} list concatenation	%{... , ...} bit concatenation
~ bitwise not	!, not boolean not
+, - unary positive, negative	*, /, % multiply, divide, modulus
+, - plus, minus	>>, << shift right, shift left
<, <=, >, >= boolean comparison	is [not] a subtype identification
==, != boolean equal, not equal	===,!== Verilog 4-state compare
~, !~ string matching	&, |, ^ bitwise and, or, xor
&&, and boolean and	||, or boolean or
!, not boolean not	=> boolean implication
a ? b : c conditional "if a then b, else c"	

A.6 Coverage Groups and Items

A.6.1 Struct and Unit Members

cover *cover-group* [using [also] *cover-group-options*] is [empty] [also] {
    item *item-name* [: *type = expr*] [using [also] *cover-item-options*];
    cross *item-name1*, *item-name2*, ... ; transition *item-name*;
};

To enable coverage, extend the global struct as follows:
    setup_test() is also {set_config(cover, mode, *cover-mode*)}

A.6.2 Coverage Group Options

text = *string*	weight = *uint*	no_collect	radix = DEC|HEX|BIN
count_only	global	when = *bool-expr*	
external=surecov		agent_options=*SureCov options*	

A.6.3   Coverage Item Options

text = *string*	when = *bool-expr*	weight = *uint*
no_collect	radix=DEC|HEX|BIN	name *name*
at_least = *num*	ignore | illegal = *cover-item-bool-expr*	
no_trace	ranges=range([*n..m*], *sub-bucket-name*, *sub-bucket-size*, *at-least-number*);	
per_instance	agent_options=*SureCov options*	

A.7 Lists

A.7.1   Selected List Pseudo-Methods

add[0](*list-item* : list-type)	add[0](*list* : list)
clear()	delete(*index* : int)
pop[0]() : list-type	push[0](*list-item* : list-type)
insert(*index* : int, *list* : list | *list-item* : list-type)	

A.7.2   Selected List Expressions

size() : int	top[0]() : list-type
reverse() : list	sort(*expr* : expr) : list
sum(*expr* : int) : int	count (*expr* : bool) : int
exists(*index* : int) : bool	has(*expr* : bool) : bool
is_empty() : bool	is_a_permutation(*list*: list) : bool
all(*expr* : bool) : list	all_indices(*expr* : bool) : list of int
first(*expr* : bool) : list-type	last(*expr* : bool) : list-type
first_index(*expr* : bool) : int	last_index(*expr* : bool) : int
key(*key-expr* : expr) : list-item	key_index(*key-expr* : expr) : int
max(*expr* : int) : list-type	max_value(*expr* : int) : int | uint
min(*expr* : int) : list-type	min_value(*expr* : int) : int | uint
swap(*small* : int, *large* : int) : list of bit	
crc_8|32(*from-byte* : int, *num-bytes* : int) : int	
unique(*expr* : expr) : list	

A.8 Temporal Language

A.8.1 Event Struct and Unit Members

event *event-name* [is [only] *TE*]; // struct or unit member

emit [*struct-inst.*]*event-name*; // action

A.8.2 Predefined Events

sys.any   *struct-inst.*quit

A.8.3 Temporal Struct and Unit Members

on *event-name* { *action*; ... } ;

expect|assume [*rule-name* is [only]] *TE*
    [else dut_error(*"string"*, *expr*, ...)];

A.8.4 Basic Temporal Expressions

@[*struct-inst.*]*event-name*   // event instance

change|fall|rise('*HDL-path*') @sim   // simulator callback annotation

change|fall|rise(*expr*)   true(*bool-expr*)   cycle

A.8.5 Boolean Temporal Expressions

TE1 and *TE2*   *TE1* or *TE2*   not *TE*   fail *TE*

A.8.6 Complex Temporal Expressions

TE @[*struct-inst.*]*event-name*	// explicit sampling
{ *TE*; *TE*; ... }	// sequence
TE1 => *TE2*	// if TE1, then TE2 follows
TE exec { *action*; ... }	// execute when TE succeeds
[*n*] [* *TE*]	// fixed repeat
{ ... ; [[*n*]..[*m*]] [* *TE*]; *TE*; ... }	// first match repeat
~[[*n*]..[*m*]] [* *TE*]	// true match repeat
delay(*expr*)	detach(*TE*)
consume(@[*struct-inst.*]*event-name*)	

A.8.7   Time Consuming Actions

wait [[until] *TE*];   sync [*TE*];

A.8.8   Using Lock and Release

```
struct struct-type {
    locker-exp: locker;
    TCM() @event-name is {
        locker-exp.lock();

        ...
        locker-exp.release();
    };
};
```

A.9 Packing and Unpacking Pseudo-Methods

expr = pack(*pack-options*, *expr*, …)
    // pack options: packing.high, packing.low

unpack(*pack-options*, *value-expr*, *target-expr* [, *target-expr*, ...])

A.10 Simulator Interface Statements and Unit Members

verilog function '*HDL-path*'(*params*) : *n*;   // n is result size in bits

verilog import *file-name*;   // statement only

verilog task '*HDL-path*'(*params*);

verilog time *Verilog-timescale*;   // statement only

vhdl driver '*HDL-path*' using *option*, …;   // unit member only

vhdl function '*designator*' using *option*, …;

vhdl procedure '*identifier*' using *option*, …;

vhdl time *VHDL-timescale*;   // statement only

A.11 Preprocessor Directives

#define [']*macro-name* [*replacement*]

#if[n]def [']*macro-name* then {*string*} [#else {*string*}] ;
Note: Preprocessor directives can be statements, struct or unit
members, or actions.

e Tidbits

Answers to commonly asked questions are provided in this appendix.

B.1 History of *e*

The history of the *e* language is rich with stories of seemingly insurmountable challenges, incredible innovations, magnificent triumphs, and an untold number of everyday successes. *Specman Elite*, and the *e* language it is based upon, represent that history in many, many ways. The strength of the *e* language for functional verification has fueled its rapid adoption and has driven the commercial success of Verisity Design, Inc. As you might expect, the histories of *e* and Verisity are intertwined and are based upon the direct verification experiences of their fore-fathers. Quite commonly, the most innovative Electronic Design Automation (EDA) tools are invented by engineers who use them for their own success, and the stories behind *e* and Verisity show how important this can be for the health of the industry.

The *e* language was invented by the founder and CTO of Verisity, Yoav Hollander. It came into being at a time that Yoav was providing verification consulting for leading edge companies in his home country of Israel. While performing this work, he was frustrated by the fact that his efforts to create a verification environment were lost when he moved to a new project. Each environment was custom developed and not reusable. By the early '80s Yoav and his team were creating innovative verification environments using random test generation and repeatable generation of tests for several of their projects. By 1992 an early version of the *e* language and a prototype of Specman Elite were in experimental use at National Semiconductor and Digital Equipment Corporation. Verisity was formed (initially as InSpec) in late 1995 in order to bring Specman Elite to market. Specman Elite was introduced to the market during the Design Automation Conference of 1996.

Yoav drew from his extensive software experience and exposure to powerful languages such as SmallTalk, Eiffel, Perl, C++, as well as the CobaltBlue and DeltaBlue constraint languages. His goals for meeting the requirements of functional verification led him to Aspect-Oriented Programming (AOP), a level of extensibility beyond Object Oriented Programming. AOP allows for the explicit extension and changing of very base building blocks of a verification environment hierarchy. This was important to enable the specification of a basic verification environment and the creation of specific tests which drove the verification environment in various unique ways. Thus AOP is one of the building block concepts of the *e* language. The building blocks of the *e* language remain the same today as at its inception: *test generation, results checking,* and *coverage measurement.*

The language itself is strongly typed, by use of a simple inheritance scheme. It is pointer free and has built in garbage collection capabilities for managing memory usage. There is a familiar regular expression syntax, as well as more comprehensive built in list types and operations. The layering scheme of the language provides the AOP abilities.

As Yoav and Verisity expanded the reach of Specman Elite to more customers, a verification community formed around the concepts of constraint-driven test generation. The importance of an expressive specification language and strong algorithm led to several generations of technological advancements. For example, the third generation constraint solver was completed in 1997. It supported multidirectional constraints, enabling the declarative nature of *e* as it is used today. This close collaboration with customers created a de facto standardization of the *e* language and the rapid adoption of Specman Elite.

This broadening of the user base led to many layers of innovation. Verisity's large research and development team extended Specman Elite's abilities, building unified behavior of methods and processes, and multithreading evaluation. Temporal language, assertions, and events were introduced for more concise expressions. Coverage capabilities were added to enable verification teams to plan and measure a design's readiness for tapeout.

By 1999 Verisity supported a rapidly growing list of third-party verification tool integrations. The demand for efficient interaction between tools led to the introduction of *units* to the *e* language and the beginning of a project to define and deliver an External Simulator Interface (ESI). The ESI makes supporting and adding simulator integrations much more efficient, and because it is a public interface, it enables outside parties to do that work.

The Verisity user community continued its rapid growth during this time, especially as the design community began to create increasingly complex System-on-Chips (SoCs), which stressed older verification technologies far beyond their inherent limitations. The sharing of intellectual property between projects and companies, which has made SOC design so successful, also introduced verification challenges which drove Verisity to define a functional verification methodology to support these advanced requirements. In 2002, this effort resulted in the *e*

Reuse Methodology, or *e*RM, culminating many staff-years of cooperative effort with customers and verification consultants. At this time Verisity also invented the concept of *e* Verification Components, or *e*VCs. An *e*VC is a verification application that is built by means of the *e*RM technology, perhaps by Verisity, by a third party, or by an end user. An *e*VC encapsulates all of the essential functional components needed to verify that a device is compliant with a protocol, including generation, checking, and functional coverage, into a single entity that can be simply added to a verification environment with a simple command.

These same trends in design complexity drove the need to support system level verification requirements. Support for hardware accelerators became crucial for Verisity's largest customers. To facilitate this, a synthesizeable subset of the *e* language was defined, which enables higher frequency verification environment constructs to be accelerated along with the design. The addition of ports was a direct result of the work to accelerate verification environments.

Now more than ever, functional verification is critical to the successful delivery of high-quality electronics. Not only have chips increased in size and complexity, but system and embedded software verification are becoming crucial for product viability. The work to innovate methodology and technology to address the seemingly intractable challenges of functional verification still continues at Verisity. The *e* language and Specman Elite continue to lead the way, enabling first-pass success for its users.

B.2 *e* Resources

B.2.1 Verisity Design, Inc.

Verisity Design, Inc., currently provides the Specman Elite simulator. Visit *http:// www.verisity.com* for updates on new products, news and company information.

B.2.2 Verification Vault

This is a rich repository of *e*-related information. It contains information on frequently asked questions, known bugs with patches and workarounds, new software releases, product documentation, and newsgroups for sharing information with other *e* users. This is a very important site for *e* users. Go to *http://www.verificationvault.com* to access the Verification Vault.

B.2.3 Club Verification

Club Verification is a users' group meeting for *e* users. It is a great place to meet with other *e* users to share ideas and insights and see how your colleagues are approaching the verification process. The conference features presentations from users on their experiences and successes with *e* and on the technical aspects of the tool and tool methodology.

For details visit *http://www.verisity.com*.

B.2.4 Specman Elite Yahoo Groups

This is a group for Specman Elite users and developers to ask questions of each other and share code, techniques, and experience as well as general discussion about Specman Elite.

Visit *http://groups.yahoo.com/group/specman/* for details.

B.3 Verification Resources

This section lists available verification resources on the World Wide Web. Although most of these websites do not directly discuss the *e* language, it is important for a verification engineer to be familiar with different verification methodologies to choose the technique that is most suitable for a specific verification problem.

B.3.1 General HDL Websites

- Verilog—*http://www.verilog.com*
- Cadence—*http://www.cadence.com/*
- EE Times—*http://www.eetimes.com*
- Synopsys—*http://www.synopsys.com/*
- DVCon (Conference for HDL and HVL Users)—*http://www.dvcon.org*
- Verification Guild—*http://www.janick.bergeron.com/guild/default.htm*
- Deep Chip—*http://www.deepchip.com*

B.3.2 Architectural Modeling Tools

- For details on SystemC, see *http://www.systemc.org*

B.3.3 Simulation Tools

- Information on *Verilog-XL* and *Verilog-NC* is available at *http://www.cadence.com*
- Information on *VCS* is available at *http://www.synopsys.com*

B.3.4 Hardware Acceleration Tools

Information on hardware acceleration tools is available at the following websites:

- *http://www.cadence.com*
- *http://www.aptix.com*
- *http://www.mentorg.com*
- *http://www.axiscorp.com*
- *http://www.tharas.com*

B.3.5 In-circuit Emulation Tools

Information on in-circuit emulation tools is available at the following websites:

- *http://www.cadence.com*
- *http://www.mentorg.com*

B.3.6 Coverage Tools

Information on coverage tools is available at the following websites:

- *http://www.verisity.com*
- *http://www.synopsys.com*

B.3.7 Assertion Checking Tools

Information on assertion checking tools is available at the following websites:

- Information on *e* is available at *http://www.verisity.com*
- Information on *Vera* is available at *http://www.open-vera.com*
- Information on *SystemVerilog* is available at *http://www.accellera.org*
- *http://www.0-in.com*
- *http://www.verplex.com*
- Information on Open Verification Library is available at *http://www.accellera.org*

B.3.8 Equivalence Checking Tools

Information on equivalence checking tools is available at the following websites:

- *http://www.verplex.com*
- *http://www.synopsys.com*

B.3.9 Formal Verification Tools

Information on formal verification tools is available at the following websites:

- *http://www.verplex.com*
- *http://www.realintent.com*
- *http://www.synopsys.com*
- *http://www.athdl.com*
- *http://www.0-in.com*

Bibliography

Manuals and References

The following manuals are available at *http://www.verificationvault.com*. These manuals provide an in-depth understanding of the topics discussed in this book.

e Language Reference Manual, Verisity Design, Inc.

Specman Elite Usage Guide, Verisity Design, Inc.

Specman Elite Tutorial, Verisity Design, Inc.

Verification Advisor Online Documentation, Verisity Design, Inc.

e Reuse Methodology (eRM) Reference Manual, Verisity Design, Inc.

Books

Peter Ashenden, *The Designer's Guide to VHDL*, Morgan Kaufmann, 2nd Edition, 2001. ISBN 1-5586067-4-2.

Lionel Bening and Harry Foster, *Principles of Verifiable RTL Design*, 2nd Edition, Kluwer Academic Publishers, 2001. ISBN 0-7923-7368-5.

Janick Bergeron, *Writing Testbenches: Functional Verification of HDL Models*, Kluwer Academic Publishers, 2000. ISBN 0-7923-7766-4.

J. Bhasker, *A Verilog HDL Primer*, Star Galaxy Publishing, 1999. ISBN 0-9650391-7-X.

J. Bhasker, *Verilog HDL Synthesis: A Practical Primer*, Star Galaxy Publishing, 1998. ISBN 0-9650391-5-3.

J. Bhasker, *A VHDL Synthesis Primer*, Star Galaxy Publishing, 2nd Edition, 1998. ISBN 0-9650391-9-6.

Michael D. Ciletti, *Modeling, Synthesis and Rapid Prototyping with the Verilog HDL*, Prentice Hall, 1999. ISBN 0-1397-7398-3.

Ben Cohen, *Real Chip Design and Verification Using Verilog and VHDL*, VhdlCohen Publishing, 2001. ISBN 0-9705394-2-8.

Ben Cohen, *VHDL Coding Styles and Methodologies*, Kluwer Academic Publishers, 1999. ISBN 0-7923847-4-1.

James M. Lee, *Verilog Quickstart*, Kluwer Academic Publishers, 1997. ISBN 0-7923992-7-7.

Samir Palnitkar, *Verilog HDL: A Guide to Digital Design and Synthesis*, 2nd Edition, Prentice Hall, 2003. ISBN 0-13-044911-3.

Douglas Smith, *HDL Chip Design: A Practical Guide for Designing, Synthesizing and Simulating ASICs and FPGAs using VHDL or Verilog*, Doone Publications, 1996. ISBN 0-9651934-3-8.

E. Sternheim, Rajvir Singh, Rajeev Madhavan, and Yatin Trivedi, *Digital Design and Synthesis with Verilog HDL*, Automata Publishing Company, 1993. ISBN 0-9627488-2-X.

Stuart Sutherland, *Verilog 2001: A Guide to the New Features of the Verilog Hardware Description Language*, Kluwer Academic Publishers, 2002. ISBN 0-7923-7568-8.

Stuart Sutherland, *The Verilog PLI Handbook: A User's Guide and Comprehensive Reference on the Verilog Programming Language Interface*, 2nd Edition, Kluwer Academic Publishers, 2002. ISBN: 0-7923-7658-7.

Donald Thomas and Phil Moorby, *The Verilog Hardware Description Language*, 4th Edition, Kluwer Academic Publishers, 1998. ISBN 0-7923-8166-1.

Sudhakar Yalamanchili, *VHDL Starters Guide*, Prentice Hall, 1997. ISBN 0-1351980-2-X.

Bob Zeidman, *Verilog Designer's Library*, Prentice Hall, 1999. ISBN 0-1308115-4-8.

Index

About the Author

Samir Palnitkar is the President of Jambo Systems, Inc., a leading ASIC design and verification services company and a Verisity Verification Alliance partner. Jambo Systems specializes in high-end designs for microprocessor, networking, and communications applications. Mr. Palnitkar previously founded Integrated Intellectual Property, Inc., an ASIC company that was acquired by Lattice Semiconductor, Inc., and Obongo, Inc., an e-commerce software firm that was acquired by AOL Time Warner, Inc.

Mr. Palnitkar holds a Bachelor of Technology in Electrical Engineering from the Indian Institute of Technology, Kanpur; a Master's in Electrical Engineering from the University of Washington, Seattle; and an MBA degree from San Jose State University, San Jose, California.

Mr. Palnitkar is a recognized authority on *e*, Verilog HDL, modeling, verification, logic synthesis, and EDA-based methodologies in digital design. He has worked extensively with design and verification on various successful microprocessor, ASIC, and system projects; worked on many *e*-based projects; and trained hundreds of students on *e* since 1997.

Mr. Palnitkar was the lead developer of the Verilog framework for the shared memory, cache coherent, multiprocessor architecture, popularly known as the UltraSPARC™ Port Architecture, defined for Sun's next generation UltraSPARC-based desktop systems. Besides the UltraSPARC CPU, he has worked on diverse design and verification projects at leading companies including Cisco, Philips, Mitsubishi, Motorola, National, Advanced Micro Devices and Standard Microsystems.

Mr. Palnitkar is the author of the best-selling book, *Verilog HDL: A Guide to Digital Design and Synthesis*, used at universities and semiconductor design companies worldwide. He is the author of three U.S. patents—one for a novel method to analyze finite state machines, a second for work on cycle simulation technology, and a third (patent pending approval) for a unique e-commerce tool. He has also published several technical papers. In his spare time, Mr. Palnitkar likes to play cricket, read books, and travel the world.

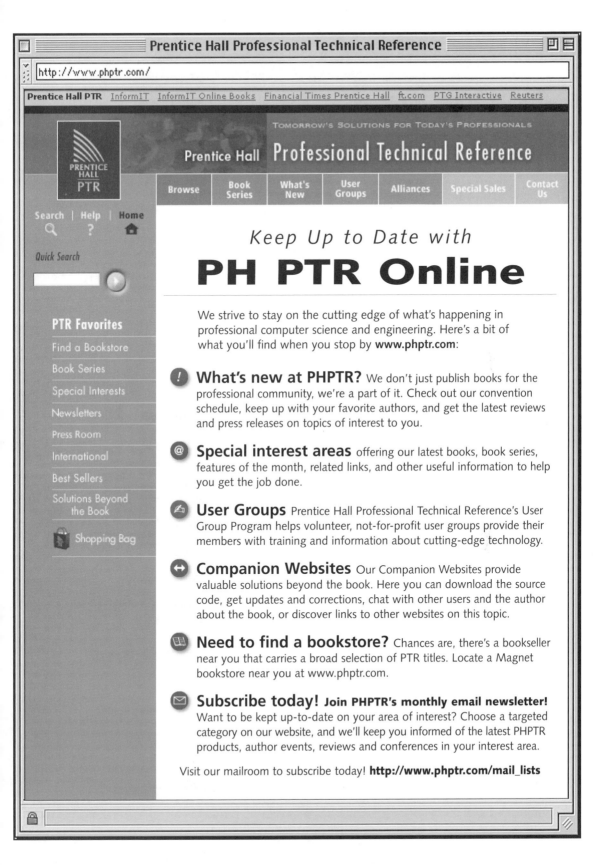

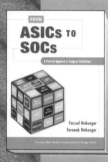
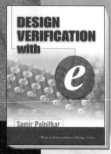

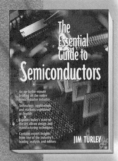
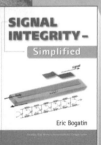